投考公務員
題解 EASY PASS
能力傾向測試

第四版

Fong Sir 著

U0152339

Common Recruitment Examination

序

公務員綜合招聘考試（Common Recruitment Examination，簡稱CRE）是特區政府聘請大部份公務員的第一關。由於通過CRE是應徵學位或專業程度公務員職位的基本條件，故應考人數年年遞增，每次均達數萬人。

政府聘請公務員，一般要求申請人需於CRE中的個別試卷中取得相當的成績，作為衡量申請人語文及推理能力的客觀標準。雖然公務員事務局表示，個別部門可以因應需要而接受未達標的申請，但申請人必須在指定日期參加CRE，並符合招聘廣告內列出的基本入職要求才會獲聘。故此，有意投考公務員空缺者，還是及早應考為上。

由於CRE的試卷並不會公開，令不少有意投身公務員行列人士難以提前做好準備工作，錯失成為公務員的機會。本書正是根據CRE最常出的題型撰寫。書中不但詳列各類題型的解題方法，更輔以練習、解說及模擬試卷，內容由淺入深，助你提升答題速度，而題目數量多達數百條，絕對是有意投考公務員空缺者的必讀之選。

本書的好處如下：

- 收錄200條題目，是備戰能力試的操卷首選。

- 深入剖析5大題型的必用概念：中文演繹推理、英文演繹推理、數據運用、數序推理及圖表分析。透過Concept teaching的方法，讓你即使面對新題種，亦能應付自如。

- 精選練習及講解，讓你熟習考試模式及出題方向。通過反覆練習，助你掌握答題技巧，提升答題速度。

- 試題講解詳盡，方便考生藉此檢視自己的強項及弱點，提升學習成效。

CONTENT 目錄

PART I 演繹推理

演繹推理旨在考察考生的邏輯推理能力：在每道題中給出一段陳述，這段陳述被假設是正確的，不容置疑的。要求考生根據這段陳述，選擇一個備選答案。

1.1 演繹推理 18 種

1. 簡單推理與直接推斷型

【定義】

這類題型的具體形式是：以題幹為前提，要求在選項中確定合乎邏輯的結論；或者，從題幹出發，不可能推出什麼樣的結論。

對一個邏輯推理，不管是簡單還是複雜，都要明確推理形式的有效性。推理形式的有效性亦稱「保真性」，指一個正確有效的推理必須確保從真的前提推出真的結論。儘管從假的前提出發也能進行合乎邏輯的推理，其結論可能是真的，也可能是假的，但從真前提出發進行有效推理，卻只能得到真結論，不會得到假結論。只有這樣，才能保證使用這種推理工具的安全性。這種保真性是對於正確推理是最起碼的要求。

其實，解決這類簡單推理或直接推斷型考題，考生只需運用日常邏輯推理就可以找到答案，幾乎沒有什麼技巧可言，這類題型中很多屬於送分題，一般可在十秒中內解決。下面對屬於此類的考題進行逐一分析。

【例題】

1. 學校複印社試行承包後，複印價格由每張標準紙0.35元上升0.40元，引起了學生的不滿。校務委員會通知承包商，或者他能確保複印的原有價格保持不變，或者將中止他的承包。承包商採取了相應的措施，既沒有因而減少了盈利，又沒有違背校務委員會通知的字面要求。

    以下哪項最可能是承包商採取的措施？

    A. 承包商會見校長，陳述因耗材（特別是複印紙）價格上漲使複印面臨難處，說服校長指令校務委員會收回通知。

    B. 承包商維持每張標準紙0.40元的複印價格不變，但由使用進價較低的A牌子複印紙改為使用進價較高的B牌子複印紙。

    C. 承包商把複印價格由每張0.40元降低為0.35元，但由使用進價較高的B牌子複印紙改為使用進價較低的A牌子複印紙。

    D. 承包商維持每張標準紙0.40元的複印價格不變，但同時增設了打字業務，其收費低於市價，受到學生歡迎。

    E. 承包商決定中止承包。

2. 古時候的一場大地震幾乎毀滅了整個人類，只有兩個部落死裡逃生。最初在這兩個部落中，神帝部落所有的人都堅信人性本惡，聖地部落所有的人都堅信人性本善，並且沒有既相信人性本善又相信人性本惡的人存在。後來兩個部落繁衍生息，信仰追隨和部落劃分也遵循著一定的規律。

部落內通婚，所生的孩子追隨父母的信仰，歸屬原來的部落；部落間通婚，所生孩子追隨母親的信仰，歸屬母親的部落。我們發現神聖子是相信人性本善的。

在以下各項對神聖子身份的判斷中，不可能為真的是：

A. 神聖子的父親是神帝部落的人
B. 神聖子的母親是神帝部落的人
C. 神聖子的父母都是聖地部落的人
D. 神聖子的母親是聖地部落的人
E. 神聖子的姥姥是聖地部落的人

3. 血液中的高濃度脂肪蛋白含量的增多，會增加人體阻止吸收過多的膽固醇的能力，從而降低血液中的膽固醇。有些人通過有規律的體育鍛煉和減肥，能明顯地增加血液中高濃度脂肪蛋白的含量。

以下哪項，作為結論從上述題幹中推出最為恰當？

A. 有些人通過有規律的體育鍛煉降低了血液中的膽固醇，則這些人一定是胖子。
B. 不經常進行體育鍛煉的人，特別是胖子，隨著年齡的增大，血液中出現高膽固醇的風險越來越大。
C. 體育鍛煉和減肥是降低血液中高膽固醇的最有效的方法。
D. 有些人可以通過有規律的體育鍛煉和減肥來降低血液中的膽固醇。
E. 標準體重的人只需要通過有規律的體育鍛煉就能降低血液中的膽固醇。

【答案及解析】

1. 答案：C
不讓漲價，又要保證自己的贏利，怎麼辦？唯有降低成本。對於本題而言，該承包商實際上是降低了服務質量，但是這並沒有違背校務委員會通知的字面要求。

2. 答案：B
B項斷定神聖子的母親是神帝部落的人，則不論神聖子的父親是哪個部落的，由題幹的條件，可推出神聖子都一定相信人性本惡。因此，B項的斷定不可能真。其餘的各項都可能真。

3. 答案：D
題幹斷定有些人通過體育鍛煉和減肥，能增加血液中的高濃度脂肪蛋白；同時，血液中的高濃度脂肪蛋白含量的增多，會降低血液中的膽固醇。由此可以推出，有些人可以通過體育鍛煉和減肥來降低血液中的膽固醇。因此，D項作為題幹的推論是恰當的。C項和D項類似，但其所做的斷定過強，作為從題幹推出的結論不恰當。其餘各項均不恰當。

2. 複雜推理與綜合推斷型

【定義】

　　此類考題表面無統一特徵，只是比直接推斷型要複雜些，當然所謂複雜，其實並不很複雜，只是要多繞些彎而已。

　　這種試題通常在題幹中給出若干條件，要求考生從這些條件中合乎邏輯推出某種結論。這類題型很多涉及複合判斷推理，特別是對假言、聯言和選言等推理的綜合運用。

　　解這類題一般一下子看不出答案，需進行深入分析和推理。其解題基本思路是，從題幹中所給條件的邏輯關係或事物的內部聯繫出發，逐步綜合進行推理。下面對屬於此類的考題進行舉例分析。

【例題】

1. 從趙、張、孫、李、周、吳六個工程技術人員中選出三位組成一個特別工程小組，集中力量研製開發公司下一步準備推出的高技術產品。為了使工作更有成效，我們了解到以下情況：

　　(1) 趙、孫兩個人中至少要選上一位

　　(2) 張、周兩個人中至少選上一位

　　(3) 孫、周兩個人中的每一個都絕對不要與張共同入選

根據以上條件，若周未被選上，則下列中哪兩位必同時入選？

A. 趙、吳

B. 張、李

C. 張、吳

D. 趙、李

E. 趙、張

2. 某市的大廈工程建設計劃正進行招標。有四個建築公司投標。為簡便起見，稱它們為公司甲、乙、丙、丁。在投標結果公佈以前，各公司經理分別做出猜測。

甲公司經理說：「我們公司最有可能中標，其他公司不可能。」

乙公司經理說：「中標的公司一定出自乙和丙兩個公司之中。」

丙公司經理說：「中標的若不是甲公司就是我們公司。」

丁公司經理說：「如果四個公司中必有一個中標，那就非我們莫屬了！」

中標結果公佈後發現，四人中只有一個人的預測成真了。以下哪項判斷最可能為真？

A. 甲公司經理猜對了，甲公司中標了

B. 乙公司經理猜對了，丙公司中標了

C. 甲公司和乙公司的經理都說錯了

D. 乙公司和丁公司的經理都說錯了

E. 甲公司和丁公司的經理都說錯了

【答案及解析】

1. 答案：E

根據條件推導即可。周未被選上，而張、周兩個人中至少選上一位，所以張肯定選上了。張選上了，孫就肯定沒選上，趙就肯定選上了。所以答案是趙、張。

2. 答案：C

我們可以對題幹中幾個公司的預言進行如下歸納：

若甲言中，則甲中標或四家都沒有中標；若甲說錯，則中標者出自乙、丙丁。

若乙言中，則中標者出自乙、丙；若乙說錯，由甲、丁中標或四家都沒中。

若丙言中，則中標者出自甲、丙；若丙說錯，則乙、丁中標或四家都沒中。

若丁言中，則丁中標或四家都沒中；若丁說錯，由甲、乙、丙中標。

對於選項C，由於甲公司和乙公司都說錯了，可以得出丁中標了。由此可知，丙錯了，只有丁說對了。因此，選項C的判斷與題幹敘述完全符合。或者說，由選項C推不出與題幹矛盾的結論。

若選項A為真，甲說對了，而且甲中標了；由此可以推出丙也說對了，與題設矛盾。若B為真，則乙和丙都猜對了，也不符合題幹假設。若D為真，則可以推出甲中標了，從而甲、丙都猜對了，與題幹不符。若選項E為真，可以推出乙或丙公司中標了。因為乙猜的是乙或丙公司中標，預測一定為真。但丙公司猜的是甲或丙中標，可能錯（若乙中標），也可能對（若丙中標）。因此，若假設選項E為真，可能推出與題幹假設矛盾的結論。

3. 直言判斷及對當關係型

【定義】

　　直言判斷及對當關係是最基本的一個邏輯知識點，試題的表現形式多樣。要解這類題型，注意的是解題時千萬不能以個人經驗或專業知識為依據，關鍵是一定要從題幹給出的內容出發，從中抽象出同屬於對當關係的邏輯形式，根據對當關係來分析判斷。

【例題】

1. 「這個部門已發現有職工違規情況。」如果上述斷定是真的，則在下述三個斷定中：

　　（1）這個部門沒有職工不違規

　　（2）這個部門有的職工沒有違規情況

　　（3）這個部門所有的職工都未有違規

　　不能確定真假的是：
　　A. 只有（1）和（2）
　　B. （1）、（2）和（3）
　　C. 只有（1）和（3）
　　D. 只有（2）
　　E. 只有（1）

2. 「你可以隨時愚弄某些人。」假若以上屬實，以下哪些判斷必然為真？

(1) 張先生和李先生隨時都可能被你愚弄

(2) 你隨時都想愚弄人

(3) 你隨時都可能愚弄人

(4) 你只能在某些時候愚弄人

(5) 你每時每刻都在愚弄人

A. 只有（3）
B. 只有（2）
C. 只有（1）和（3）
D. 只有（2）、（3）和（4）
E. 只有（1）、（3）和（5）

【答案及解析】

1. 答案：A

選項（1）的意思是這個部門所有的職工都有違規情況了。這跟題幹不矛盾，有可能是真的，但也可能是假的，不能確定真假。選項（2）的意思是這個部門有的職工沒有違規。這跟題幹也不矛盾，不能確定真假。選項（3）的意思是這個部門沒有一個職工違規，這與題幹矛盾，肯定為假。

2. 答案：A

我們先解釋（3）為真。題幹說「你可以隨時愚弄某些人」，是指「你經常具有愚弄某些人」的「能力」，其中「某些人」是一個確定的集合，比如李小姐、王先生。（3）說的是「你隨時都可能愚弄人」，意思是「你愚弄人」的行為「隨時」都可能發生。（3）並沒有關心被愚弄的人是誰，只要有一個人就行，是上述確定集合的人當然可以。因此，由題幹（具有能力）可以推出（3）（行為可能發生）。

其次，我們說明除（3）以外，其他不是必然的結論。關於（1），請注意張先生和李先生可能不屬於「你」要以愚弄的人的集合。（2）說的是「隨時」在「想」愚弄人，與題幹意思不合，因為題幹只是說「可以」隨時愚弄人，但並不一定每時每刻都在愚弄人，也不一定每時每刻都在「想」愚弄人。（5）說「你只能在某些時候愚弄人」，劃定了一部分時間，與題幹說的「隨時」不符。（5）斷言「你每時每刻都在愚弄人」，強調了愚弄人的行為總在發生，與題幹所指的「能力總存在」是不一樣的。

4. 加強支持型

【定義】

在邏輯考試中，圍繞前提和結論之間的支持或反駁關係，設計了多種形式的考題，主要有加強前提型和削弱結論型。

前提對結論的支持或反駁程度，有許多推理或論證儘管不滿足保真性，即前提的真不能確保結論的真，但前提卻對結論提供一定程度的支持，或者前提對結論構成一定程度的反駁。一個推理的證據支持度越高，則在前提真實的條件下，推出的結論可靠性越大。

在批判性思維以及邏輯考試中，重點考察的就是思維的論證性，即對各種已有的推理或論證做批判性評價：

－ 對某個論點是否給出了理由？

－ 所給出的理由真實嗎？

－ 與所要論證的論點相關嗎？

－ 如果相關，對論點的支持度有多高？

－ 是必然性支持（若理由真，則論點或結論必真），還是或然性支持（若理由真，結論很可能真，也可能是假）？

－ 是強支持還是弱支持？

－給出什麼樣的理由能夠更好地支持該結論？

－給出什麼樣的理由能夠有力地駁倒該結論，或者至少是削弱它？

加強支持型考題的題型特點從問題的問法上就可以看出來，這種題型的問法主要有：

（1）增加以下哪項，則肯定有***的結論？

（2）以下哪項如果為真，則最能支持（加強）以上結論？

（3）以下哪項如果為真，則最能支持（加強）***的結論？

（4）支持型和削弱型的變種：除***之外，都加強（或削弱）

加強支持型考題解題思路是，要注意尋找與題幹一致的選項。而如果是最不能加強型，當然與題幹相矛盾或不一致的選項就最不能加強了。應該說，加強支持型和削弱質疑型是密切相關的，不論加強還是削弱，題幹的選項都必須首先與題幹相關，緊扣題幹，與題幹不相干、不一致的選項都不能加強題幹，也不能削弱題幹。下面屬於此類的考題進行舉例分析。

【例題】

1. 過度工作和壓力都會不可避免地導致失眠。森達公司的所有管理人員都有壓力。儘管醫生反覆提出警告，但大多數的管理人員每周工作仍然超過六十小時，而其餘的管理人員每周僅工作四十小時。只有每周工作超過四十小時的員工才能得到一定的獎金。

   以上的陳述最強地支持下列哪項結論？

   A. 大多數得到一定獎金的森達公司管理人員患有失眠症
   B. 森達公司員工的大部分獎金給了管理人員
   C. 森達公司管理人員比任何別的員工組更易患失眠症
   D. 沒有每周僅僅工作四十小時的管理人員工作過度
   E. 森達公司的工作比其它公司的工作壓力大

2. 甲報章每年都要按期公佈當年國產和進口電視機銷量的排名榜。專家認為這個排名榜不應該成為每個消費者決定購買哪種電視機的基礎。

   以下哪項，最能支持這個專家的觀點？

   A. 購買甲報章的人不限於要購買電視機的人
   B. 在甲報章上排名較前的電視機製造商利用此舉進行廣告宣傳，以吸引更多的消費者。
   C. 每年的排名變化較小
   D. 對任何兩個消費者而言，可以根據自己具體情況的不同，而有不同的購物標準。

E. 一些消費者對他們根據甲報章上的排名所購買的電視機很滿意。

3. 體內不產生「P450」物質的人與產生P450物質的人比較，前者患腦退化症的可能性三倍於後者。因為P450物質可保護腦部組織不受有毒化學物質的侵害。因此，有毒化學物質可能導致帕金遜式綜合症。

下列哪項，如果為真，將最有力地支持以上論證？

A. 除了保護腦部不受有毒化學物質的侵害，P450對腦部無其它作用。

B. 體內不能產生P450物質的人，也缺乏產生某些其他物質的能力。

C. 一些腦退化症病人有自然產生P450的能力

D. 當用多已胺（一種腦部自然產生的化學物質）治療腦退化症病人時，病人的症狀會得到減輕。

E. 很快就有可能合成P450，用以治療體內不能產生這種物質的病人。

4. 一則政府廣告勸告人們：酒後不要駕駛車輛，直到你感到能安全駕駛的時候才開。然而，在醫院進行的一項研究中，酒後立即被詢問的對象往往低估他們恢復駕駛能力所需要的時間。這個結果表明，在駕駛前飲酒的人很難遵循這個廣告的勸告。

下列哪項，如果為真，最強地支持以上結論？

A. 對於許多人來說，如果他們計劃飲酒的話，他們會事先安排不飲酒的人開車送他們回家。

B. 醫院中被研究的對象估計他們能力，通常比其它飲酒的人更保守。

C. 一些不得不開車回家的人就不飲酒。

D. 醫院研究的對象也被詢問，恢復對安全駕駛不起重要作用的能力所需要的時間。

E. 一般的人，對公共事業廣告的警覺比醫院研究對象的警覺高。

【答案及解析】

1. 答案：A

得到一定獎金的森達公司的管理人員肯定每周工作超過四十小時，即所謂過度工作，再加上題幹中説「森達公司的所有管理人員都有壓力」，而「過度工作和壓力都會不可避免地導致失眠症」，所以我們綜合得出了「大多數得到一定獎金的森達公司管理人員患有失眠症」這個結論，即選項A。

選項B、D討論的是管理人員以外的員工，選項C、E又引入了其它公司的情況，這都是我們不能斷定的信息，題幹無從支持。

2. 答案：D

選項E是反對這個專家的。選項D支持了這個專家的觀點。

3. 答案：A

這個推理的主線是：（1）P450可以保護腦部組織不受有毒化學物質的侵害，（2）體內產生P450物質的人比較不容易患腦退化症。所以，有毒化學物質可能導致腦退化症。這個推理其實不能説完備，有P450的人身上有兩種性質，要想證明這兩種性質之間的因果關聯，那需要保證沒有別的一些性質在中間折騰。選項A「除了保護腦部不受有毒化學物質的侵害，P450對腦部無其它作用」，就滿足這一點。如果沒有選項A，比如，除了保護腦部不受有毒化學物質的侵害之外，P450還能夠產生一種什麼抗原，這種抗原能夠使人腦免受腦退化症之擾的話，題幹的推理就出問題，不成立了。

4. 答案：B

選項B中的「保守」這個詞的含義理解很重要。選項B説明其他飲酒的人可能會比醫院中的那些研究對象更冒險，也就是更加低估他們恢復駕駛能力所需要的時間。這對題幹的結論是個很有力的支持。

5. 削弱質疑型

【定義】

削弱質疑型是邏輯考試的一個重點，歸結為此類題型的考題是數量是最大的。削弱質疑型考題的主要問法有：

（1）以下哪項如果為真，則能最嚴重地削弱（反駁）以上結論？

（2）以下哪項如果為真，則最能削弱（反駁）***的結論？（要小心）

（3）在上述結論中，***忽略了哪一種可能性？

（4）以下哪項如果為真，則指出了上述論證的邏輯錯誤？

（5）下列哪項如果為真，則對以上結論提出最嚴重的質疑？

（6）支持型和削弱型的變種：除***之外，都加強（或削弱）

要想削弱這種說法，就要強調如果不做這件事情，會有很壞的負效果。比如有人說考試不用溫習邏輯題，因為他溫習了之後才得了30分，如你要反駁對方，就得說，你如果不溫習的話，連30分都得不了。

　　削弱題型的解題關鍵是首先應明確原文的推理關係，即什麼是前提、什麼是結論；在此基礎上，尋找削弱的基本方向是針對前提、結論還是論證本身。具體對不同的情況有不同的處理，比如：

　　類型一，直接反對原因，即直接說明原文推理的前提不正確，就達到推翻結論的目的；

　　類型二，指出存在其他可能解釋，原文以一個事實、研究、發現或一系列數據為前提推出一個解釋上述事實或數據的結論，要削弱這個結論，就可以通過指出由其他可能來解釋原文事實；

　　類型三，原文認為A不是導致B的原因，要對其進行削弱，就可以指出A是B的間接原因，即指出A通過導致C而間接的導致了B。

　　下面對如何削弱，做一個詳細分析。首先要明確，要使一個結論為真，必須滿足兩個條件：①前提真實，②推理或論證形式有效。於是要反駁或削弱某個結論，通常有這樣幾條途徑：一是直接反駁結論，其途徑有：舉出與該結論相反的一些事實（舉反例），或從真實的原理出發構造一個推理或論證，以推出該結論的否定；二是反駁論據，即反駁推出該結論的理由和根據；三是指出該推理或論證不合邏輯，即從前提到結論的過渡是不合法的，違反邏輯規則。

　　如果是削弱結論，那麼首先要搞清題幹中的結論是什麼；

如果是反對什麼觀點，特別要注意的是問題問的是反對的是誰的觀點，什麼觀點；如果對推理提出質疑，那麼就要搞清題幹的推理結構和前提條件是什麼，一個有效的推理必須前提成立，推理形式正確，才能得出正確的結論。這類題目要求在選項中確定哪一項為真，能構成對題幹中論證的一個反駁，從而也就削弱了該論證的結論。另外，在削弱結論型考題中，有時雖然要確立的選項不直接構成對論題、論據或推理形式的反駁，但作為前提加入到題幹的原前提中去以後，會減低證據支持度，減低結論的可靠性，從而削弱題幹的論證。

如果是削弱論證，那麼一定要搞清其實是要削弱什麼？這就要求我們要對題幹部分的論證進行盡可能的簡化，抓住中間最主要的推理關係。解題思路是尋找一種弱化的方式，使其既可以是肯定選項中與題幹的結論不相容的選項，也可以從選項中找到一個使題幹的論證不能成立的條件。要構成對題幹中的推理的一個反駁，歸謬法是一種有效方法，具體就是舉出另一個推理，它有同樣的形式並且有真實的前提，卻得出了假的結論。

需要指出的是，對於削弱題型還要分清最能削弱型還是最不能削弱型。如果是最不能削弱型，解題時應先將能削弱題幹的與題幹唱反調的選項排除掉，最後剩下的選項與題幹不相干還是支持題幹的都是最不能削弱的。如果是最能削弱型，則應首先將選項與題幹一致的選項排除掉，同時尋找與題幹相矛盾或不一致的選項，從中進一步比較削弱的程度。

【例題】

1. 據對一批企業的調查顯示，這些企業總經理的平均年齡是
    57歲，而在20年前，同樣的這些企業的總經理的平均年齡
    大約是49歲。這說明，目前企業中總經理的年齡呈老化趨
    勢。

    以下哪項，對題幹的論證提出的質疑最為有力？

    A. 題幹中沒有說明，20年前這些企業關於總經理人選
       是否有年齡限制。
    B. 題幹中沒有說明，這些總經理任職的平均年數。
    C. 題幹中的信息，僅僅基於有20年以上歷史的企業。
    D. 20年前這些企業的總經理的平均年齡，僅是個近似
       數字。
    E. 題幹中沒有說明被調查企業的規模。

2. 一項全球範圍的調查顯示，近10年來：吸煙者的總數基本保持不變；每年只有10%的吸煙者改變自己的品牌，即放棄原有的品牌而改吸其他品牌；煙草製造商用在廣告上的支出佔其收入的10%。

   在Z煙草公司的年終董事會上，董事A認為，上述統計表明，煙草業在廣告上的收益正好等於其支出，因此，此類廣告完全可以不做。董事B認為，由於上述10%的吸煙者所改吸的香煙品牌中，幾乎不包括本公司的品牌，因此，本公司的廣告開支實際上是筆虧損性開支。以下哪項，構成對董事A的結論的最有力質疑？

   A. 董事A的結論忽視了：對廣告開支的有說服力的計算方法，應該計算其佔整個開支的百分比，而不應該計算其佔收入的百分比。

   B. 董事A的結論忽視了：近年來各種品牌的香煙的價格有了很大的變動。

   C. 董事A的結論基於一個錯誤的假設：每個吸煙者在某個時候只喜歡一種品牌。

   D. 董事A的結論基於一個錯誤的假設：每個煙草製造商只生產一種品牌。

   E. 董事A的結論忽視了：世界煙草業是一個由處於競爭狀態的眾多經濟實體組成的。

3. 美國法律規定，不論是駕駛員還是乘客，坐在行駛的小汽車中必須綁好安全帶。有人對此持反對意見。他們的理由是：每個人都有權冒自己願意承擔的風險，只要這種風險不會給別人帶來損害。因此，坐在汽車裡不戴好安全帶，純粹是個人的私事，正如有人願意承擔風險去炒股，有人願意承擔風險去攀岩純屬他個人的私事一樣。以下哪項，如果為真，最能對上述反對意見提出質疑？

A. 儘管確實為了保護每個乘客自己，而並非為了防備傷害他人，但所有航空公司仍然要求每個乘客在飛機起飛和降落時綁好安全帶。

B. 汽車保險費近年來連續上漲，原因之一，是由於不繫安全帶造成的傷亡使得汽車保險賠償費連年上漲。

C. 在實施了強制要求綁安全帶的法律以後，美國的汽車交通事故死亡率明顯下降。

D. 法律的實施帶有強制性，不管它的反對意見看來多麼有理。

E. 炒股或攀岩之類的風險是有價值的風險，不綁安全帶的風險是無謂的風險。

【答案及解析】

1. 答案：C

正如C項所指出的，題幹的論據，僅僅基於有20年以上歷史的老企業。而題幹的結論，卻是對包括新老企業在內的目前各種企業的一般性評價。如果上述這樣的老企業在目前的企業中佔的比例不大，則題幹結論的可信度就會大為降低。因此，C項是對題幹的有力質疑。其餘各項均不能構成對題幹的質疑。

2. 答案：E

題幹中統計的煙草業的廣告收益，是世界各煙草企業收益的合計；同樣，這樣的廣告支出，也是世界各煙草企業支出的合計。因此，雖然從合計上看收支相當，但對單個的煙草企業來說，其在廣告上的支出和收益可以表現出很大的差別。因此顯然不能因為題幹的數據顯示煙草業在廣告上的收益等於其支出，而得出此類廣告完全可以不做的結論。董事A的觀點正是忽視了這樣一個事實，即世界煙草業是一個由處於競爭狀態的眾多經濟實體組成的。E項指明了這一點，因而構成對董事A的結論的有力質疑。其餘各項對董事A的質疑均不得要領。

3. 答案：B

如果B項為真，則說明不戴安全帶不是汽車主人的純個人私事，它引起的汽車保險費的上漲損害了全體汽車主的利益。這就對題幹中的反對意見提出了有力的質疑。其餘各項均不能構成有力的質疑。

6. 傳遞排序型

【定義】

　　傳遞排序型題目，一般在題幹部分給出不同對象之間的若干個兩兩對比的結果，要求從中推出具體的排序。解答這類題型的主要思路是要把所給條件抽像成最簡單的排序形式。

【例題】

1. 有A、B、C、D四個有實力的排球隊進行循環賽（每個隊與其他隊各比賽一場），比賽結果B隊輸掉一場，C隊比B隊少贏一場，而B隊又比D隊少贏一場。

　　關於A隊的名次，下列哪一項為真？
A. 第一名
B. 第二名
C. 第三名
D. 第四名
E. 條件不足，不能斷定。

2. 有四個外表看起來沒有分別的小球，它們的重量可能有所不同。取一個天秤，將甲、乙歸為一組，丙、丁歸為另一組分別放在天秤的兩邊，天秤是基本平衡的。將乙和丁對調一下，甲、丁一邊明顯地要比乙、丙一邊重得多。可奇怪的是，我們在天秤一邊放上甲、丙，而另一邊剛放上乙，還沒有來得及放上丁時，天秤就壓向了乙一邊。請你判斷，這四個球中由重到輕的順序是什麼？

    A. 丁、乙、甲、丙。
    B. 丁、乙、丙、甲。
    C. 乙、丙、丁、甲。
    D. 乙、甲、下、丙。
    E. 乙、丁、甲、丙。

【答案及解析】

1. 答案：D

四個隊單循環，每個隊要賽三場。根據題幹，「B隊輸掉一場，C隊比B隊少贏一場」，可得出C隊輸兩場，「而B隊又比D隊少贏一場」，得出D隊贏三場。也就是D第一名，A全輸，第四名，就是倒數第一。

2. 答案：A

根據題幹已知信息可知：

（1）甲+乙=丙+丁；將乙和丁對調後：

（2）甲+丁＞丙+乙；由在天秤一邊放上甲、丙，而另一邊則放上乙：

（3）甲+丙＜乙；

由（1）+（2）可得：甲+乙+甲+丁＞丙+丙+乙+丁，即甲＞丙；

由（1）-（2）可得：丙+丁+甲+丁＞甲+乙+丙+乙，即丁＞乙；

由（3）可得：乙＞甲；綜合以上可知：丁＞乙＞甲＞丙，即丁球重於甲球。因此正確答案為A項。

7. 計算統計及數字陷阱型

【定義】

計算統計及數字陷阱型考題出現的也不少，因此，有必要介紹一下有關的統計常識。

統計結論的可靠性主要取決於樣本的代表性。只有從能夠代表總體的樣本出發，才能得到關於總體的可靠結論。一般從抽樣的規模、抽樣的廣度和抽樣的隨機性三個方面去保證樣本的代表性。對於任何一個抽樣統計結果，我們都可以從這些角度去質疑它的可靠性。

在當代社會，各種數字、數據、報表可以說鋪天蓋地，頻頻出現在大眾傳媒之中，我們常常會想這些數字、數據準確、可靠嗎？對「精確」數字保持必要的警惕，應該說是一種明智的、理性的態度。下面揭示一些隱藏在「精確」數字背後的陷阱：

一是平均數陷阱，在對平均數的模糊理解做文章；

二是百分比陷阱，一般題幹僅提供兩種事物的某種比率就比較出兩種事物的結果，其實其陷阱就在於該百分比所賴以計算出來的基數是不同的；

三是錯誤比較，或者不設定供比較的對象，不設定比較的根據或基礎。

因此，表面上在進行比較，實際上根本就不能比較。

同時，我們還應該懂一些計量經濟學研究的基本觀念：不能只看到兩組數據之間的正比或反比關係，關鍵要分析其背後的社會經濟聯繫。比如兩組數據的相關性很強（存在正比或反比關係），但僅此而已相互間並沒有任何因果關係，有時也可能兩組數據都是由第三種數據決定的。

【例題】

1. 在過去的十年中，由美國半導體工業生產的半導體增加了200%，但日本半導體工業生產的半導體增加了500%，因此，日本現在比美國制造的半導體多。以下哪項為真，最能削弱以上命題？

   A. 在過去五年中，由美國半導體工業生產的半導體增長僅100%。
   B. 過去十年中，美國生產的半導體的美元價值比日本生產的高。
   C. 今天美國半導體出口在整個出口產品中所佔的比例比十年前高。
   D. 十年前，美國生產的半導體佔世界半導體的90%，而日本僅2%。
   E. 十年前，日本生產半導體是世界第4位，而美國列第一位。

2. 據國際衛生與保健組織1999年年會「通訊與健康」公佈的
   調查報告顯示，68%的腦癌患者都有經常使用流動電話的
   歷史。這充分說明，經常使用流動電話將會極大地增加一
   個人患腦癌的可能性。

   以下哪項如果為真，則將最嚴重地削弱上述結論？

   A. 進入20世紀80年代以來，使用流動電話者的比例有
      驚人的增長。
   B. 有經常使用流動電話的歷史的人在1990至1999年超
      過世界總人口的65%。
   C. 在1999年全世界經常使用流動電話的人數比1998年
      增加了68%。
   D. 使用普通電話與流動電話通話者同樣有導致癌的危
      險。
   E. 沒有使用過流動電話的人數在90年代超過世界總人
      口的50%。

3. 世界衛生組織在全球範圍內進行了一項有關捐血對健康影響的跟蹤調查。調查對象分為三組：第一組對象中均有兩次以上的捐血記錄，其中最多的達數十次；第二組中的對象均僅有一次捐血記錄；第三組對象均從未捐過血。

調查結果顯示，被調查對象中癌症和心臟病的發病率，第一組分別為0.3%和0.5%，第二組分別為0.7%和0.9%，第三組分別為1.2%和2.7%。一些專家依此得出結論，捐血有利於減少患癌症和心臟病的風險。這兩種病已經不僅在發達國家，而且也在發展中國家成為威脅中老人生命的主要殺手。因此，捐血利己利人，一舉兩得。以下哪項如果為真，將削弱以上結論？

（1）60歲以上的調查對象，在第一組中佔60%，在第二組中佔70%，在第三組中佔80%。

（2）獻血者在獻血前要經過嚴格的體檢，一般具有較好的體質。

（3）調查對象的人數，第一組為1700人，第二組為3000人，第三組為7000人。

A. 只有（1）
B. 只有（2）
C. 只有（3）
D. 只有（1）和（2）
E. （1）、（2）和（3）

【答案及解析】

1. 答案：D

按選項D和題幹中的數據估算，「現在」美國半導體的產量仍為日本的20多倍，削弱了題幹的結論「日本現在比美國製造的半導體多」。如果真算一下，90% (1+200%)/ [2%(1+500%)] = 270/12 = 22.5。

選A不對。五年增100%比十年增200%所説的年增長速度更快；選項B、C與題幹所説的無直接關係；E雖有削弱題幹命題的意思，但名次的差別不能準確反映其數量之間的差異，從削弱題幹的命題的角度看明顯不如D。

2. 答案：B

如果一份對中國人的調查顯示，肺癌患者中90%以上都是漢族人，由此顯然不能得出結論，漢族人更容易患肺癌，因為漢族人本身就佔了中國人的90%以上。同樣的道理，如果B項的斷定為真，説明在世界總人口中，有經常使用流動電話歷史的人所佔的比例，已接近在腦癌患者中有經常使用流動電話歷史的人所佔的比例，這就嚴重削弱了題幹的結論。其餘各項均不能削弱題幹的結論。

3. 答案：D

（1）能削弱題幹的結論。因為在三個組中，60歲以上的被調查對象呈10%遞增，又題幹斷定，癌症和心臟病是威脅中老人生命的主要殺手，因此，有理由認為，三個組的癌症和心臟病發病率的遞增，與其中中老年人比例的遞增有關，而並非説明捐血有利於減少患癌症和心臟病的風險。

（2）能削弱題幹的結論。因為如果捐血者一般有較好的體質，則捐血記錄較高的調查對象，一般患癌症和心臟病的可能性就較小，因此，並非是捐血減少了他們患癌症和心臟病的風險。

（3）不能削弱題幹。因為題幹中進行比較的數據是百分比，被比較各組的絕對人數的一定差別，不影響這種比較的説服力。

8. 真話假話型

【定義】

　　真話假話型涉及了邏輯基本規律（同一律、矛盾律、排中律），解決這類問題的突破口，往往是運用對當關係等邏輯知識在所有敘述中找出有互相矛盾的判斷，從而必知其一真一假。

　　例如下列兩個命題是互相矛盾的：

　　（1）「所有S是P」與「有些S不是P」

　　（2）「所有S不是P」與「有些S是P」

　　（3）「p並且q」與「或者p或者非q」

　　（4）「p或者q」與「非p並且非q」

　　（5）「如果p則q」與「p並且非q」

　　（6）「只有p才q」與「非p並且q」

　　（7）「必然p」與「可能非p」

　　（8）「不可能p」與「可能p」

　　要注意的是：有時兩個命題雖然不是矛盾的，但互相反對（或下反對），即不能同真（或不能同假），那就可以推出兩個判斷中至少有一個是假的（或者至少有一個是真的），這也同樣是解題的關鍵。

例如下列命題是互相反對的（不能同真，但可以同假）：

（1）「所有S是P」與「所有S不是P」

（2）「所有S都是P」與「（某個）S不是P」

（3）「所有S不是P」與「（某個）S是P」

（4）「必然p」與「不可能（必然非）p」

例如下列命題是互相下反對的（不能同假，但可以同真）：

（1）「有些S是P」與「有些S不是P」

（2）「有些S是P」與「（某個）S不是P」

（3）「有些S不是P」與「（某個）S是P」

（4）「可能p」與「可能非p」

真話假話型考題包括一真多假、一假多真或多真多假三種，主要問法有：

（1）上述判斷中只有一個為真（假），以下哪項最可能為真？

（2）這幾句話中只有兩句是真，請你推出以下哪項為真？

解題基本思路主要有兩種：一是用矛盾（或反對）法，具體做法是從題幹提供的所有判斷中，找到兩個矛盾（或反對）的判斷，從而知其真假關係，進一步可推理出答案；二是用假設反證法，這種方法雖然顯得笨些，卻很有實效。

【例題】

1. 桌子上有4個杯，每個杯上面都寫著一句話：

   第一個杯：「所有的杯子中都有水果糖」

   第二個杯：「本杯中有蘋果」

   第三個杯：「本杯中沒有巧克力」

   第四個杯：「有些杯子中沒有水果糖」

   如果其中只有一句真話，那麼以下哪項為真？

   A. 所有的杯子中都有水果糖
   B. 所有的杯子中都沒有水果糖
   C. 所有的杯子中都沒有蘋果
   D. 第三個杯子中有巧克力
   E. 第二個杯子裡有蘋果

2. 某班有一位同學做了好事沒留下姓名，他是甲、乙、丙、
   丁四人中的一個。當老師問他們時，他們分別這樣說：

   甲：這件好事不是我做的。

   乙：這件好事是丁做的。

   丙：這件好事是乙做的。

   丁：這件好事不是我做的。

   這四人中只有一個人說了真話，請你推出是誰做了好事？
   A. 甲
   B. 乙
   C. 丙
   D. 丁
   E. 不能推出

3. 某律師事務所共有12名工作人員，其中：

(1) 有人會使用電腦；

(2) 有人不會使用電腦；

(3) 所長不會使用電腦。

上述三個判斷中只有一個是真的。

以下哪項正確表示了該律師事務所會使用電腦的人數？

A. 12人都會使用

B. 12人沒人會使用

C. 僅有一個不會使用

D. 僅有一人會使用

E. 不能確定

【答案及解析】

1. 答案:D

題幹中第一和第四個杯上的話是矛盾的,兩句話中必有且只有一句為真。因此,四句中的一句真話必在第一和第四之中,所以第二和第三個杯上的話必為假。由第三個杯子上的話「本杯中沒有巧克力」是假,可知D中所說「第三個杯子中有巧克力」為真。

雖然第二個杯上的話也假,但五個選項中沒有一項是「第二個杯沒有蘋果」。選項C是「所有的杯中都沒有蘋果」,不等同於「第二個杯沒有蘋果」,因為,前者能推出後者,後者不能推出前者。

若選A,則第一個和第三個杯上的話為真。若選E,則第二個和第四個杯上的話為真。選B或選C也不對。我們用以下例子說明這兩種選擇是錯誤的。設第一個杯裝水果糖,第二個杯裝水果糖,第三個杯裝巧克力,第四個杯裝蘋果。對這一「分配方案」,只有第四個杯上的話是對的,符合題幹的假設,但B和C不真。雖然可以舉出符合題幹而B和C又為真的例子,但此例說明,由題幹不能推出B和C必然為真。

2. 答案:A

簡捷做法:觀察發現乙和丁兩個人說的話相互矛盾的,肯定當中有一個人說的話為真,而四人中只有一個人說了真話,這樣,甲和丙說的話就肯定是假的。從甲的話為假,可知好事是甲做的,而且這仍然保持丙的話為假。

3. 答案:A

用假設反證法:如果(3)為真,即「所長不會使用電腦」,那就是「有人不會使用電腦」,即(2)也肯定為真,與題幹的假設「只有一個是真的」矛盾,因此,(3)肯定應該為假,即所長會用電腦。由此可推出「有人會用電腦」,即(1)為真,進一步可推出(2)為假,即沒有人不會用電腦,也就是律師事務所12名工作人員都會用電腦。

9. 假言命題及推理型

【定義】

假言命題及推理型考題，主要是考察充分條件和必要條件的區分及具體運用，這是邏輯考試中一個常考的題目類型。

要解這類題型，首先要搞清充分條件和必要條件，並要根據常見的連接詞能迅速抽取出邏輯形式。充分條件和必要條件在中文中的代表詞分別是只要和只有。

比如：「只要A就B」意思是「A是B的充分條件，從A可以推出B，從B不一定推出A」；「只有A才B」意思是「A是B的必要條件，從B可以推出A，從A不一定推出B」。

做這類題型要求能熟練掌握假言判斷與假言推理的運用，具體要熟悉：

（1）推理的傳遞性（A推出B B推出C則A能推出C）

（2）不可逆性（重要的考點，A推出B B真，推不出A真）

（3）逆否命題（A推出B則非B推出非A）

解決這類題型的基本思路是一般可用直接推理方式解決，比如：「如A做，則B一定做；若C做則B 不做；於是，若A做則C不

做（否則矛盾）」。如果已知條件很多很亂的問題時，要迅速找到答案有一定的難度，因此，要同時考慮已知條件和選項，在理解了已知條件的基礎上迅速瀏覽選項，從兩頭推理，從而盡快找到答案。

【例題】

1. 「如果缺乏奮鬥精神，就不可能有較大成就。李先生有很強的奮鬥精神，因此，他一定能成功。」

    下述哪項為真，則上文推論可靠？

    A. 李先生的奮鬥精神異乎尋常
    B. 不奮鬥，成功只是水中之月
    C. 成功者都有一番奮鬥的經歷
    D. 奮鬥精神是成功的唯一要素
    E. 成功者的奮鬥是成功的前提

2. 當爸爸、媽媽中只有一個人外出時，兒子可以留在家裡。如果爸爸、媽媽都外出，必須找一個保姆，才可以把兒子留在家中。

    從上面的陳述中，可以推出下面哪項結論？

    A. 兒子在家時，爸爸也在家。
    B. 兒子在家時，爸爸不在家。
    C. 保姆不在家，兒子不會單獨在家。
    D. 爸爸、媽媽都不在家，則兒子也不在家。
    E. 爸爸不在家，則媽媽在家。

3.　「如果你犯了法，你就會受到法律制裁；如果你受到法律制裁，別人就會看不起你；如果別人看不起你，你就無法受到尊重；而只有得到別人的尊重，你才能過得舒心。」

從上述敘述中，可以推出下列哪一個結論？

A. 你不犯法，日子就會過得舒心。

B. 你犯了法，日子就不會過得舒心。

C. 你日子過得不舒心，證明你犯了法。

D. 你日子過得舒心，表明你看得起別人。

E. 如果別人看得起你，你日子就能舒心。

【答案及解析】

1. 答案：D

題幹第一句話「如果缺乏奮鬥精神，就不可能有較大成就」告訴我們「奮鬥精神」是「較大成就」的必要條件。而題幹第二句又從「李先生有很強的奮鬥精神」，直接推出「他一定能成功」的結論，就要求把「奮鬥精神」作為「成功」的充分條件。如果奮鬥精神是成功的唯一要素，成功的必要條件就是充分條件。因此，加上D中給的條件後，題幹中的推論就能成立。

而選項B、C、E只是以不同方式重複題幹中第一句話的意思。選項A雖然能對題幹中的推理有增強作用，但只是程度上的增強，仍不能得出「一定能成功」的必然結論，因為李先生可能並不具備影響成功的其他要素，比如「機遇」。

2. 答案：C

選項C中這個「單獨在家」是解題的關鍵。保姆不在家，肯定爸爸和媽媽中至少一個人留在家裡陪兒子呢。選項A、B、D、E都不一定。

3. 答案：B

從題幹順推，我們能夠得到選項B：你犯了法，日子就不會過得舒心。

我們來看選項C：你日子過得不舒心，證明你犯了法。這不一定。日子過得不舒心，還可能是病了呢，還可能是沒錢花呢。「不犯法」只是「日子過得舒心」的必要條件。

選項A「你不犯法，日子就會過得舒心」，這又把「不犯法」當成「日子過得舒心」的充分條件了。

選項D不著邊際。

選項E也把「別人看得起你」當成了「日子過得舒心」的充分條件了，這不對，應該是必要條件，別人看得起我，可是我就是找不到女朋友，日子也舒心不了。

10. 集合或範圍重合型

【定義】

集合或重合型題型一般特點：在題目中出現「所有」、「有些」、「某個」、「每一個」、「沒有一個」等集合型的敘述或題幹提供的概念間的範圍有重合的部分。

可以根據基本的集合概念和邏輯常識解決該類題型，解這種題型的重點放在集合的「部分與全體」上，同時要善於分辨可能重合的部分和絕不會重合的部分。最直觀的辦法是根據題幹提供的條件畫個小圖，題目即可迎刃而解。

【例題】

1. 在某住宅小區的居民中，大多數中老年教師都買了人壽保險，所有買了四居室以上住房的居民都買了財產保險，而所有辦了人壽保險的都沒有買財產保險。如果上述斷定是真的，以下哪項關於該小區居民的斷定必定是真的？

(1) 有中老年教員買了四居室以上的新房。

(2) 有中老年教員沒辦理財產保險。

(3) 買了四居室以上住房的居民都沒辦理人壽保險。

    A.（1）、（2）和（3）

    B. 僅（1）和（2）

    C. 僅（2）和（3）

D. 僅（1）和（3）

E. 僅（2）

2. 根據剛才的題幹，如果在題幹的斷定中再增加以下斷定：
「所有的中老年教師都買了人壽保險」，並假設這些斷定
都是真的，那麼，以下哪項必定是假的？

A. 在買了四居室以上住房的居民中有中老年教師

B. 並非所有辦理人壽保險的都是中老年教師

C. 某些中老年教師沒買四居室以上的住房

D. 所有的中老年教師都沒有買財產保險

E. 某些買了人壽保險的沒買四居室以上的住房

第3和第4題基於以下題幹。

本問題發生在一所學校內：學校的教授們中有一些是足球
迷。學校的預算委員會的成員們一致要把學校的足球場改建為一
個科貿寫字樓，以改善學校收入狀況。所有的足球迷都反對將學
校的足球場改建成科貿寫字樓。

3. 如果以上各句陳述均為真，則下列哪項也必為真？

   A. 學校所有的教授都是學校預算委員會的成員
   B. 學校有的教授不是學校預算委員會的成員
   C. 學校預算委員會有的成員是足球迷
   D. 並不是所有的學校預算委員會的成員都是學校的教授
   E. 有的足球迷是學校預算委員會的成員

4. 如果作為上面陳述的補充，明確以下條件：所有的學校教授都是足球迷，那麼下列哪項一定不可能是真的？

   A. 有的學校教授不是學校預算委員會的成員
   B. 學校預算委員會的成員中有的是學校教授
   C. 並不是所有的足球迷都是學校教授
   D. 所有的學校教授都反對將學校的足球場改建為科貿寫字樓
   E. 有的足球迷不是學校預算委員會的成員

【答案及解析】

1. 答案：C

大多數中老年教師都買了人壽保險，而所有買了人壽保險的居民都沒有買財產保險，所以大多數中老年教師沒買財產保險，這是（2）。

買了四居室以上住房的居民都買了財產保險，而所有買了人壽保險的居民都沒辦理財產保險，所以，買了四居室以上住房的居民都沒有買人壽保險（否則矛盾了），這是（3）。中老年教師和四居室以上住房之間沒有建立因果聯繫，推不出（1）來。

2. 答案：A

題幹發生了一個變化，從「大多數中老年教師都買了人壽保險」加強為「所有的中老年教師都買了人壽保險」，這就意味著所有的中老年教師都沒有買財產保險。而買了四居室以上住房的居民都買了財產保險，這就是說「所有的中老年教師」和「買了四居室以上住房的居民」這兩個集合沒有任何交集。所以這選項A「在買了四居室以上住房的居民中有中老年教師」就為假了。

3. 答案：B

B項必為真。因為所有的足球迷都反對改建足球場，而所有的預算委員會的成員都主張改建足球場，因此，所有的預算委員會的成員都不是足球迷。又有的教授是足球迷。因此有的教授不是預算委員會的成員。其餘各項均不必定為真。

4. 答案：B

B項不可能真。因為所有的教授都是足球迷，因此所有的教授都反對改建足球場；而所有的預算委員會的成員都主張改建足球場。因此不可能有預算委員會的成員是教授。

11. 預設或尋找假設型

【定義】

預設或尋找假設型也是考試的一個重點。這類題型主要表現形式有：

（1）加上一個條件就變成了一個有效的三段論推理，比如題目中問到「上面的邏輯前提是哪個？」「再加上什麼條件能夠得出結論」；

（2）題幹給出前提和結論，然後提問你假設是什麼？或者要你提出正面的事實或有利於假設的說明，則加強論點，否則削弱論點。比如問到：

－「上文的說法基於以下那一個假設？」

－「上述結論中隱含著下列哪項假設？」

－「上述議論中假設了下列哪項前提？」

（3）以題幹為結論，要求在選項中確定能推出題幹的前提。或者，需要補充什麼樣的前提，才能使題幹中的推理成為邏輯上有效的推理？

由於這種題型是題幹推理中的前提不足夠充分以推出結論，要求在選項中確定合適的前提，去補充的原前提或論據，從而能

合乎邏輯地推出結論或有利於提高推理的證據支持度和結論的可靠性。因此，做這類題的基本思路是緊扣結論，簡化推理過程，從因果關係上考慮，從前提到結論，中間一定有適當的假設，尋找斷路或是因為「顯然」而省略掉的論述，也就是要「搭橋」，很多時候憑語感或常識就可以找到所要問的隱含的前提。

解題還可以從這幾個方面考慮：

一是，所尋找的假設應當是原文命題成立的必要條件；

二是，假設的正確選項如果取反，能夠推翻原文的推理；

三是，對不確定選項的判斷取反後，看是否能夠推翻原文。

進一步説，這類題型涉及到邏輯中的「預設」問題。預設有語義預設和語用預設之分，語義預設是一個命題及其否定都要假定的東西，是一個命題能夠為真或為假的前提條件；語用預設則可以表述為，如果話語A只有當命題B為交談雙方所共知時才是恰當的，則A在語用上預設B。由於日常交際中的推理都是具有某些共同背景知識的人在特定的語境中進行的，不必列出所有必需的前提，但要注意的是這種省略本身可能不是真的或這種省略推理中可能暗含著邏輯錯誤。因此，在批判性思維中，常常需要把這些被省略的前提、假定、預設補充到推理過程中來，以便考察被省略的前提是否真實，推理過程是否正確，即對推理者的推理進

行評價。

同時，由於這種省略形式的前提對結論提供了不充分的支持，有時候需要加強前提以便對結論提供更強的支持，或對該論證提供更好的辯護。被補充到前提中去的，可以是某個一般原理如因果關係陳述，也可以是某個假設、假定或事實性斷言。對前提的加強可以到使該推理成為形式有效的推理的地步，但更多的時候只是提高了推理中前提對結論的證據支持度。

【例題】

1. 母親要求兒子從小就努力學外語。兒子說：「我長大又不想當翻譯，何必學外語。」

   以下哪項是兒子的回答中包含的前提？

   A. 要當翻譯需要學外語
   B. 只有當翻譯才需要學外語
   C. 當翻譯沒什麼大意思
   D. 學了外語才能當翻譯
   E. 學了外語也不見得能當翻譯

2. 無論是商業用電還是家居用電，現行的電費價格一直偏低。某區推出一項舉措，對超出月額定數的用電量，無論是商業用電還是家用電，一律按上調高價收費。這一舉措將對該區的節約用電產生重大的促進作用。上述舉措要達到預期的目的，以下哪項必須是真的？

（1）有相當數量的浪費用電是因為電價格偏低而造成的。

（2）有相當數量的用戶是因為電價格偏低而浪費用電的。

（3）超額用電價格的上調幅度一般地足以對浪費用電的用戶產生經濟壓力。

A.（1）、（2）和（3）
B. 僅（1）和（2）
C. 僅（1）和（3）
D. 僅（2）和（3）
E.（1）、（2）和（3）都不必須是真的

3. 在西方幾個核武大國中，當核試驗得到了有效的限制，老百姓就會傾向於省更多的錢，出現所謂的商品負超常消費；當核試驗的次數增多的時候，老百姓就會傾向於花更多的錢，出現所謂的商品正超常消費。因此，當核戰爭成為能普遍覺察到的現實威脅時，老百姓為存錢而限制消費的願望大大降低，商品正超常消費的可能性大大增加。

上述論證基於以下哪項假設？

A. 當核試驗次數增多時，有足夠的商品支持正超常消費。

B. 在西方幾個核大國中，核試驗受到了老百姓普遍地反對。

C. 老百姓只能通過本國的核試驗的次數來覺察核戰爭的現實威脅。

D. 商界對核試驗乃至核戰爭的現實威脅持歡迎態度，因為這將帶來經濟利益。

E. 在冷戰年代，上述核戰爭的現實威脅出現過數次。

【答案及解析】

1. 答案：B

兒子的結論是不學外語，理由是不想當翻譯，其推理過程是：「只有當翻譯，才要學外語，我不想當翻譯，所以不要學外語」（這是一個必要條件推理，把「當翻譯」作為「學外語」的必要條件）。

選項C只是反映兒子對當翻譯的態度，排除。選項E是說「學外語」不是「當翻譯」的充分條件，但並不能說「當翻譯」是「學外語」的必要條

件，不選。選項A、D不選，因為該兩項選擇表示「當翻譯」是「學外語」的充分條件，並不一定必要，不當翻譯也可能需要學外語。

2. 答案：C

容易把「有相當數量的用戶是因為電價格偏低而浪費用電的」這個選項也選進來，其實不妥。比如儘管相當數量的用戶因為電價低而浪費用電，但浪費的總量很有限，在浪費用電的總量中有相當數量的浪費用電是因為公費支出的（所以大家浪費不在乎）或是缺乏節電意識造成的。那麼你提高電價就不一定達到目的，而不如採取別的措施（比如限制公費繳納電費的總額或是加強宣傳節電的力度等等）了。

3. 答案：A

題幹的論證必須基於這一假設，即當核試驗次數增多時，有足夠的商品支持正超常消費。否則，即使老百姓傾向於花更多的錢，也不會出現商品正超常消費。因此，A項是必須假設的。

題幹說明，老百姓能通過本國的核試驗的次數來覺察核戰爭的現實威脅，但顯然並不說明，老百姓只能通過本國的核試驗的次數來覺察核戰爭的現實威脅。因此，C項並非是必須假設的。

D項是對題幹論證的一個可能的推論，並非是題幹的論證本身必須假設的。

E項是對過去的一個陳述，不是必須假設的。

但細究一下，本題還是存在漏洞，因為本題沒有明確「正超常消費」的確切含義，如果核試驗增多時，雖然沒有足夠的商品，但老百姓有更強的消費欲，引起通貨膨脹，物價上漲，還算不算「有足夠的商品支持正超常消費」。從這個角度看，選項A也未必是必須要假設的。

12. 說明解釋型

【定義】

　　說明解釋型考題也是一種重要的題型，其主要表現形式是，在題幹中給出某種需要說明、解釋的現象，再問什麼樣的理由、根據、原因能夠最好地解釋該現象，或最不能解釋該現象，即與該現象的發生不相干。

具體問法有：

（1）下列哪項最能解釋上述現象？

（2）以下那種說法能夠解釋上文的數據或結果？

（3）以下那種說法能夠有助於解釋上文敘述中的矛盾？

　　解這類題型有時需要一些相關的背景知識，但這些知識都屬於語言常識和一般性常識，並且已經在題幹或選項中給出，只是要求從中做一些選擇和判斷而已。需要特別指出的是，對解釋結果型，一般首先要明確解釋的關鍵概念；對解釋矛盾或差異型，思考的關鍵是抓住原文差異雙方的差別，正是這一差別導致了現象的矛盾。

【例題】

1. 「全國各地的航空公司目前開始為旅行者提供網上訂票服務。然而,在近期內,電話訂票並不會因此減少。」除了以下哪項外,其他各項均有助於解釋上述現象?

    A. 正值當地人外遊旺季,需要訂票的數量劇增。

    B. 儘管已經過技術測試,這種新的網上訂票系統距離正式運作,還需進一步測試。

    C. 絕大多數通過電話訂票的旅行者還沒有條件使用網絡。

    D. 在網上訂票系統的試用期內,大多數旅行者為了保險起見願意選擇電話訂票。

    E. 網上訂票服務的成本大大低於電話訂票,而且還有更多的選擇。

2. 英國研究各類精神緊張症的專家們發現,越來越多的人在使用互聯網之後都會出現不同程度的不適反應。根據一項對10,000名經常上網的人的抽樣調查,承認上網後感到煩躁和惱火的人數達到了三分之一;而20歲以下的網迷則有百分之四十四承認上網後感到緊張和煩躁。有關心理專家認為確實存在著某種「互聯網狂躁症」。

    根據上述資料,以下哪項最不可能成為導致「互聯網狂躁症」的病因?

    A. 由於上網者的人數劇增,通道擁擠,如果要訪問比較繁忙的網址,有時需要等待很長時間。

    B. 上網者經常是在不知道網址的情況下搜尋所需的資料和信息,成功的概率很小,有時花費了工夫也得

不到預想的結果。

C. 雖然在有些國家使用互聯網是免費的，但在我國實行上網交費制，這對網絡用戶的上網時間起到了制約作用。

D. 在互聯網上能夠接觸到各種各樣的信息，但很多時候信息過量會使人們無所適從，失去自信，個人注意力喪失。

E. 由於匿名的緣故，上網者經常會受到其他一些上網者的無禮對待或接收到一些莫名其妙的信息垃圾。

3. 有以下兩個事實：

事實1：電視廣告已經變得不是那麼有效：在電視上推廣的品牌中，觀看者能夠回憶起來的比重在慢慢下降。

事實2：電視的收看者對由一系列連續播出的廣告組成的廣告段中第一個和最後一個商業廣告的回憶效果，遠遠比對中間的廣告的回憶效果好。

以下哪項如果為真，事實2最有可能解釋事實1？

A. 由於互聯網的迅速發展，人們每天用來看電視的平均時間減少了。

B. 為了吸引更多的觀眾，每個廣告的總時間長度減少了。

C. 一般電視觀眾目前能夠記住的電視廣告的品牌名稱，還不到他看過的一半。

D. 在每一小時的電視節目中，廣告段的數目增加了。

E. 一個廣告段中所包含的電視廣告的平均數目增加了。

【答案及解析】

1. 答案：E
網上訂票服務的成本低，又有更多的選擇，是可能減少電話訂票的。

2. 答案：C
「互聯網狂躁症」是由於經常上網引起的。在我國實行上網交費制，會減少用戶的上網時間，所以不會成為「互聯網狂躁症」的病因。

3. 答案：E
題幹的事實2斷定，在一段連續插播的電視廣告中，觀眾印象較深的是第一個和最後一個，其餘的則印象較淺；而E項斷定，一個廣告段中所包含的電視廣告的平均數目增加了。

由以上兩個條件可推知，近年來，在觀眾所看到的電視廣告中，印象較深的所佔的比例逐漸減少，這就從一個角度合理地解釋了，為什麼在電視廣告所推出的各種商品中，觀眾能夠記住其品牌名稱的商品的比重在下降。其餘各項都不能起到上述作用。其中，A項可能有利於說明，隨著互聯網的迅速發展，人們所看的電視廣告的數量減少，但不能說明，在人們所看過的電視廣告中，為什麼能記住的比重降低。

13. 閱讀理解或語義分析型

【定義】

　　閱讀理解或語義分析型考題在邏輯考試中也比較常見。在我們平時的語言表達中也往往存在邏輯問題。對於日常語言由於所處環境的不同以及受話人個體的差異，往往有不同的理解。但在特定的語境下，一句話的含義應該是確定的。在需要確定一句話或一段話的真實含義時，有必要進行一定的語義分析。

　　具體說，我們通常進行推理時，前提和結論之間總是存在著某種共同的意義內容，使得我們可以由前提推出結論。形式邏輯通常不理會推理內容的相關性，但批判性思維和以它為基礎的邏輯考試卻要顧及前提和結論之間的這種內容相關性，並為此設計了許多要考慮題幹和備選答案之間的語義關聯的考題。

　　因此，解這類題的基本思路：

　　一是要閱讀仔細，通過對選項和題幹的內容逐一對照，從迅速發現找到答案的線索；

　　二是，充分運用自己平時積累起來的語感，力求準確理解、分析和推斷題幹給出的日常語言表達的句子或內容的複雜含義和深層意義。

【例題】

1. 任何方法都是有缺陷的。如何公正合理選撥合格的大學生？目前通行的高考制度恐怕是所有帶缺陷的方法中最好的方法了。以下各項都符合上述斷定的含義，除了：

   A. 被錄取的大多數大學生的實際水平與他們的考分是基本相符的。

   B. 存在落榜的考生，他們有較高的實際水平。

   C. 存在被錄取的考生，他們並無合格的實際水平。

   D. 目前，沒有比高考更能使人滿意的招生制度。

   E. 無合格的實際水平的考生被錄取，是考場舞弊所致。

2. 據《科學日報》消息，1998年5月，瑞典科學家在有關領域的研究中首次提出，一種對防治腦退化症有特殊功效的微量元素，只有在未經加工的加勒比椰果中才能提取。

   如果《科學日報》的上述消息是真實的，那麼，以下哪項不可能是真實的？

   （1）1997年4月，芬蘭科學家在相關領域的研究中提出過，對防治腦退化症有特殊功效的微量元素，除了未經加工的加勒比椰果，不可能在其他對象中提取。

   （2）荷蘭科學家在相關領域的研究中證明，在未經加工的加勒比椰果中，並不能提取對防治腦退化症有特殊功效的微量元素，這種微量元素可以在某些深海微生物中提取。

   （3）著名的蘇格蘭醫生查理博士在相關的研究領域中證

PART I 演繹推理

明，該微量元素對防治腦退化症並沒有特殊功效。

A. 只有（1）
B. 只有（2）
C. 只有（3）
D. 只有（2）和（3）
E. （1）、（2）和（3）

【答案及解析】

1. 答案：E
選項E將「無合格的實際水平的考生被錄取」歸結為「考場舞弊」，完全忽略了高考制度本身的缺陷，與題幹的斷定不符。

2. 答案：A
（1）不可能是真實的。因為由題幹，上述觀點，是瑞典科學家在1998年5月首次提出的，因此，芬蘭科學家不可能在1997年4月已經提出過。
（2）和（3）都可能是真的。因為題幹只是斷定，《科學日報》登載的消息是真實的，而沒有斷定消息中提到的瑞典科學家的觀點是真實的。

14. 匹配及邏輯運算型

【定義】

　　匹配及邏輯運算型題型一般特點是，這類題型題幹一般提供幾類因素，每類因素又有幾種不同情況，同時題幹還給出屬於不同類因素之間不同情況的判斷，要求推出確定的結論。

　　有的考生特別害怕這種匹配類型的題目，但其實只要細心得法，這類題目並不難。解這類考題時，所要使用的推理形式和推理步驟較多，推理過程顯得相對複雜。

　　解題基本思路是：通過對題幹給出的多種因素間的關係進行分析推理和排列組合，弄清題幹中所給條件的內在關係，從一個一個條件出發，逐步推理，直至推出正確答案。具體比如可以用假設反證法，耐心點推是個笨辦法，但絕對是個好辦法；也可以用表格法，把已知條件劃在一個表格上，再進一步推理。

【例題】

1. 有甲、乙、丙三個學生，一個出生在B市，一個出生在S市，一個出生在W市。他們的專業，一個是金融，一是管理，一個是外語。已知：

　　（1）乙不是學外語的

（2）乙不出生在W市

（3）丙不出生在B市

（4）學習金融的不出生在S市

（5）學習外語的出生在B市

根據上述條件，可推出甲所學的專業是：

A. 金融

B. 管理

C. 外語

D. 金融或管理

E. 推不出來

2. 方先生、王先生和余先生，一個是江西人，一個是安徽人，一個是上海人。余先生的年齡比上海人大，方先生和安徽人不同歲數，安徽人比王先生年齡小。

根據上述斷定，以下結論都不可能推出，除了：

A. 方先生是江西人，王先生是安徽人，余先生是上海人。

B. 方先生是安徽人，王先生是江西人，余先生是上海人。

C. 方先生是安徽人，王先生是上海人，余先生是江西人。

D. 方先生是上海人，王先生是江西人，余先生是安徽人。

E. 方先生是江西人，王先生是上海人，余先生是安徽人。

【答案及解析】

1. 答案：C

已知乙不是學外語的。那麼乙或者是金融或者是管理。設乙＝金融，則乙不出生在S市（因為：學習金融的不出生在S市）；所以，乙只能出生在B市（因為：乙也不出生在W市）。然而學習外語的出生在B市，因此，矛盾，假設不成立。結論：乙＝管理。丙不出生在B市，同時，學習外語的出生在B市，所以，甲＝外語。

2. 答案：D

容易檢驗當D為真時，與題幹的論述不矛盾。同時，我們也可以從題幹推出D。

由「余先生的年齡比上海人大」與「安徽人比王先生年齡小」，可以得出江西人最大；由「方先生和安徽人不同歲，安徽人比王先生年齡小」可知余先生是安徽人；

由「余先生的年齡比上海人大」可知上海人最小；由「安徽人比王先生年齡小」可知王宜是江西人。

剩下沒有判定的方先生只能是上海人。與選項D完全一致。既然選項D是可以從題幹嚴格推理得出的，其他選項與D又不同，所以其他選擇都是錯的。

15. 因果關係及因果倒置型

【定義】

因果聯繫是世界萬物之間普遍聯繫的一個方面，科學研究的一個重要任務就是要把握事物之間的因果聯繫，以便掌握事物發生、發展的規律。因果關係的主要特點有：一是普遍必然性，指任何現象都有其因，也有其果，且同因必同果，但同果卻不一定同因；二是共存性，指原因和結果總是共同變化的；三是先後性，即所謂的先因後果，但先後關係並不等於因果關係；四是複雜多樣性，指因果聯繫是多種多樣的，固然有「一因一果」，但更多的時候是「多因一果」。

排除歸納法是求因果聯繫的一個常用方法，其基本思路是：考察被研究現象出現的一些場合，在它的先行現象或恆常伴隨的現象中去尋找它的可能的原因，然後有選擇地安排某些事例或實驗，根據因果關係的上述特點，排除一些不相干的現象或假設，最後得到比較可靠的結論。

因果關係及因果倒置型在邏輯考試中出現的形式有多種，比如，為了檢查的某種因果關係是否為真，最可靠的實驗方法是改變原因後，看結果是否不同，即進行對比實驗，對比實驗的關鍵是讓實驗對象的其他方面的條件相同。又比如，有時兩組數據之

間的數據因果並不一定有原理因果，可能兩組數據都是由其它某一種數據決定的，這就是所謂表面因果與事實因果不符。

【例題】

1. 有人認為雞蛋黃的黃色跟雞所吃的綠色植物性飼料有關，為了驗證這個結論，下面哪種實驗方法最可靠？

   A. 選擇一優良品種的蛋雞進行實驗
   B. 化驗比較植物性飼料和非植物性飼料的營養成分
   C. 選擇品種等級完全相同的蛋雞，一半喂食植物性飼料，一半喂食非植物性飼料。
   D. 對同一批蛋雞逐漸增加（或減少）植物性飼料的比例。
   E. 選出不同品種的蛋雞，喂同樣的植物性飼料。

2. 某學院最近進行了一項有關獎學金對學習效率是否有促進作用的調查，結果表明：獲得獎學金的學生比那些沒有獲得獎學金的學生的學習效率平均要高出25%。調查的內容包括自習的出勤率、完成作業所需要的平均時間、日平均閱讀量等許多指標。這充分說明，獎學金對幫助學生提高學習效率的作用是很明顯的。

   以下哪項，如果為真，最能削弱以上的論證？

   A. 獲得獎學金通常是因為那些同學有好的學習習慣和高的學習效率。

B. 獲得獎學金的同學可以更容易改善學習環境來提高學習效率。

C. 學習效率低的同學通常學習時間長而缺少正常的休息。

D. 對學習效率的高低跟獎學金的多少的關係的研究應當採取定量方法進行。

E. 沒有獲得獎學金的同學普遍覺得學習壓力過重，很難提高學習效率。

3. 越來越多的有說服力的統計數據表明，具有某種性格特徵的人易患高血壓，而另一種性格特徵的人易患心臟病，如此等等。因此，隨著對性格特徵的進一步分類研究，通過主動修正行為和調整性格特徵以達到防治疾病的可能性將大大提高。

以下哪項，最能反駁上述觀點？

A. 一個人可能會患有與各種不同性格特徵均有關係的多種疾病。

B. 某種性格與其相關的疾病可能由相同的生理因素導致。

C. 某一種性格特徵與某一種疾病的聯繫可能只是數據上的巧合，並不具有一般性意義。

D. 人們往往是在病情已難以扭轉的情況下，才願意修正自己的行為，但已為時太晚。

E. 用心理手段醫治與性格特徵相關的疾病的研究，導致心理療法遭到淘汰。

【答案及解析】

1. 答案：C
對比實驗的關鍵是讓實驗對象的其他方面的條件相同：選項C「選擇品種等級完全相同的蛋雞」正是為了保證這一點（比如蛋雞的品種、年齡等）。
A沒談何種實驗；
B只講化驗，而從飼料到雞蛋還會有複雜的轉化過程，不是僅靠化驗能解決的；
D只是變化飼料的量，對驗證題幹的結論來說，不及C中的方法可靠；
E不對，因題幹中的結論並不是對某些特定的雞而言的。

2. 答案：A
因果倒置。不是因為得獎學金所以學習效率才高，而是因為學習效率高才得獎學金。

3. 答案：B
題幹中根據統計發現：甲現象（某性格特徵）總伴隨著乙現象（某疾病）出現，因此推斷，甲是乙的原因。
由B項，甲和乙可能是丙（某種生理因素）的共同結果。這就有力地反駁了題幹中甲和乙存在因果關係（可通過調整性格來治病）的觀點

16. 邏輯錯誤型

【定義】

　　邏輯錯誤型考題較多地出現在早期的邏輯考試中，近來有減少的趨勢。考對邏輯錯誤的辨析，多考邏輯錯誤的類比，比如問你「題幹中所犯邏輯錯誤與下列備選項中的哪一項最為類似？」也就是讓考生比較題幹和選項中所犯邏輯錯誤的相同或不同。這類邏輯錯誤類比型考題本書已把其歸到下一節「推理形式或結構類比型」。這裡只把邏輯考試中直接判斷邏輯錯誤型的考題歸到本節。

【例題】

1. 「平反是對處理錯誤的案件進行糾正」。以下哪項最為確切地說明上述定義的不嚴格？

   A. 對案件是否處理錯誤，應該有明確的標準。

   B. 應該說明平反的操作程序

   C. 應該說明平反的主體及其權威性

   D. 對平反的客體應該具體分析。平反了，不等於沒錯誤。

   E. 對原來重罪輕判的案件進行糾正不應該稱為平反。

2. 東方日出，西方日落，社會是發展的，生物是進化的，都反映了不依人的意志為轉移的客觀規律。王先生對此不以為然。他說，有的規律是可以改造的。人能改造一切，當然也能改造某些客觀規律。比如價值規律不是乖乖地為精明的經營者服務了嗎？人不是把肆虐的洪水制住而變害為利了嗎？

試問，以下哪項最為確切地提示了王先生上述議論中的錯誤？

A. 他過高地估計了人的力量
B. 他認為「人能改造一切」是武斷的
C. 他混淆了「運用」與「改造」這兩個概念
D. 洪水並沒有都被徹底制服
E. 價值規律若被改造就不叫價值規律了

【答案及解析】

1. 答案：E
處理錯誤的案件包括重罪輕判、輕罪重判和無罪而判。對後兩種案件進行糾正都可以叫作平反，而對於第一種進行糾正，不能叫作平反。

2. 答案：C
且不說洪水是不是被人類徹底制服，這不是關鍵，關鍵是小王認為肆虐的洪水被人們制住而變害為利是改造規律的體現，其實這是人類運用規律的體現。精明的經營者也只是掌握了價值規律然後為我所用，怎麼可能改造價值規律呢？價值規律若能被人隨意改造那就不叫「規律」了（注意：這與選項E並不同）。所以，選C。

17. 推理形式或結構類比型

【定義】

形式比較型考題是主要從形式結構上比較題幹和五個選項之間的相同或不同，即比較幾個不同推理在結構上的相同或者不同。提問形式有：

（1）「下面哪一個選項與題幹中的……最為類似？」

（2）「下面哪一個選項與題幹中的……不同？」

（3）「下面的各選項都與題幹中的……相同，除了」

（4）「下面各選項都與題幹中的不同，除了」，等等。

其解題基本思路是，著重考慮從具體的、有內容的思維過程的論述中抽像出一般形式結構，即用命題變項表示其中的單個命題，或用詞項變項表示直言命題中的詞項，每一個推理中相同的命題或詞項用相同的變項表示，不同的命題或詞項用不同的變項表示。

做這類題型只考慮推理結構和形式，而不考慮其內容的對錯，一種出題方式就是題幹本身的推理是錯誤，來對你造成一定的思維困難。

【例題】

1. 一切有利於生產力發展的方針政策都是符合人民根本利益的，改革開放有利於生產力的發展，所以改革開放是符合人民根本利益的。

以下哪種推理方式與上面的這段論述最為相似？

A. 一切行動聽指揮是一支隊伍能夠戰無不勝的紀律保證。所以，一個企業、一個地區要發展，必須提倡令行禁止、服從大局。

B. 經過對最近六個月銷售的健身器跟蹤調查，沒有發現一台因質量問題而退貨或返修。因此，可以説這批健身器的質量是合格的。

C. 如果某種產品超過了市場需求，就可能出現滯銷現象。「卓群」領帶的供應量大大超過了市場需求，因此，一定會出現滯銷現象。

D. 凡是超越代理人權限所簽的合同都是無效的。這份房地產建設合同是超越代理權限簽定的，所以它是無效的。

E. 我們對一部分實行產權明晰化的企業進行調查，發現企業通過明晰產權都提高了經濟效益，沒有發現反例。因此我們認為，凡是實行產權明晰化的企業都能提高經濟效益。

2. 法制的健全或者執政者強有力的社會控制能力，是維持一個國家社會穩定的必不可少的條件。Y國社會穩定但法制尚不健全。因此，Y國的執政者具有強有力的社會控制能力。

以下哪項論證方式，和題幹的最為類似？

A. 一個影視作品，要想有高的收視率或票房價值，作品本身的質量和必要的包裝宣傳缺一不可。電影《青樓月》上映以來票房價值不佳但實際上質量堪稱上乘。因此，看來它缺少必要的廣告宣傳和媒介炒作。

B. 必須有超常業績或者30年以上服務於本公司的工齡的僱員，才有資格獲得X公司本年度的特殊津貼。黃先生獲得了本年度的特殊津貼但在本公司僅供職5年，因此他一定有超常業績。

C. 如果既經營無方又鋪張浪費，則一個個將嚴重虧損。Z公司雖經營無方但並沒有嚴重虧損，這說明它至少沒有鋪張浪費。

D. 一個罪犯要實施犯罪，必須既有作案動機，又有作案時間，在某案中，W先生有作案動機但無作案時間。因此，W先生不是該案的作案者。

E. 一個論證不能成立，+或者它的論據虛假，或者它的推理錯誤。J女士在科學年會上關於她的發現之科學價值的論證儘管邏輯嚴密，推理無誤，但還是被認定不能成立。因此，她的論證中至少有部分論據虛假。

【答案及解析】

1. 答案：D

題幹的推理結構是：所有M都是P；S是M。所以S是P。在各選項中，只有D項具有和題幹相同的結構。

2. 答案：B

題幹的論證結構是：只有P或者Q，才R。R並且非P。因此，Q。

A項的結構是：只有P並且Q，才R。非R並且P。因此，非Q。

B項的結構是：只有P或者Q，才R。R並且非Q。因此，P。

C項的結構是：如果P並且Q，則R。P並且非R。因此，非Q。

D項的結構是：只有P並且Q，才R。P並且非Q。因此，非R。

E項的結構是：R，當且僅當P或者Q。非Q並且R。因此，P。

顯然，在各選項中，B項和題幹的結構最為類似。

18. 確定論點及繼續推論型

【定義】

　　確定論點及繼續推論型題型的具體表現形式是給出一段文字或對話，要求總結它們所表達的中心內容是什麼或什麼內容沒在題幹中表達。或給出一段論述，要求推出結論（確定論點型暨繼續推論型的變種：我們不可能得出的結論是）。

　　解題基本思路是對語言的理解，解此類題型主要是要憑語感、常識和日常的邏輯推理能力去尋找隱含的結論或內在的含義。

【例題】

1.　朱先生與王先生在討論有關用手習慣的問題。

朱：在當今85至90歲的人中，你很難找到左撇子。

王：在70年前，小孩用左手吃飯和寫字就要挨打，所以被迫改用右手。

王先生對朱先生的回答能夠加強下面哪個論斷？

　　A. 天生的右撇子有生存優勢，所以長壽。

　　B. 用手習慣是遺傳優勢與社會壓力的共同產物。

　　C. 逼迫一個人改變用手習慣是可以辦到的，也是無害的。

D. 在過去的不同時代人們對用左手還是用右手存在著不同的社會態度。

E. 小時候養成的良好習慣可以受用終生，而小時候的不良習慣也會影響終生。

2.　在大型遊樂場公園裡，現場表演是刻意用來引導人群流動的。午餐時間的表演是為了減輕公園餐館的壓力；傍晚時間的表演則有一個完全不同的目的：鼓勵參觀者留下來吃晚餐。表面上不同時間的表演有不同的目的，但這背後，卻有一個統一的潛在目標，以下哪一選項作為本段短文的結束語最為恰當？

A. 盡可能地減少各遊覽點的排隊人數。

B. 吸引更多的人來看現場表演，以增加利潤。

C. 最大限度地避免由於遊客出入公園而引起交通阻塞。

D. 在盡可能多的時間裡最大限度地發揮餐館的作用。

E. 盡可能地招徠顧客，希望他們再次來公園遊覽。

【答案及解析】

1. 答案：B

從王先生的話中可以得出兩點：一是，70年前小孩中有左撇子，這是遺傳造成的；二是，這些左撇子被迫習慣改用右手，是挨打等外在的壓力造成的。這就說明，用手習慣是遺傳優勢與社會壓力的共同產物。因此，小王的話加強了B項的斷定。

而王先生的話說明逼迫一個人改變用手習慣是可以辦到的，但並沒有說明這種改變是無害的。因此C項不成立。其餘選項均不成立。

2. 答案：D

由題幹知，大型遊樂場公園裡的有兩個經營項目：現場表演與公園餐館。從題幹的陳述不難發現，第一個項目是為第二個項目服務的，即現場表演的目的，是通過對人群流動的引導，在盡可能多的時間裡最大限度地發揮餐館的作用。因此，D項恰當。其餘各項所斷定的都可能是現場表演的，作為結束語言均不恰當。

1.2 快速答題技巧

1. 「2-3-1」的閱讀技巧

一道標準的片段閱讀試題可分為三部分：

（1）正文

（2）問題

（3）選項

按照一般的閱讀理解題的解題方式，我們會習慣性地按照「1-2-3」這樣的順序閱讀，而這樣一種閱讀方式也一般人的正常閱讀習慣。可這恰恰陷入了一種誤區。

公務員考試的特殊性質已決定了試題的「非常規」性，即要求應考者能「迅速而又準確地理解文字材料內涵的能力」。這種要求在一般性的常規考試中體現得不太明顯。可是「迅速」二字道破天機，那麼「2-1-3」模式正是我們需要使用的解題方式：先看問題部分，在頭腦中快速反應出題的考察點（是歸納主旨還是細節的判斷等），然後根據考察點回到語段中閱讀理解尋找答案，從而加快完成做題的時間。

【例題】

　　儘管國際上對此存在很多爭議，意大利文化部門還是決定用蒸餾水清洗米開朗基羅的曠世傑作——《大衛》雕像。這項工程的目的之一是除去這座高達4.5米的雕像上的塵土和油污。佛羅倫斯博物館的負責人說：「這項工程並不是為了讓雕像變得更好看。」意大利文化部長已經排除了清洗的可能性。

1.　「國際上對此存在很多爭議」的「此」指的是：
　　A.《大衛》雕像是米開朗基羅的曠世傑作
　　B. 這項工程是為了讓雕像變得更好看
　　C. 決定用蒸餾水清洗《大衛》雕像
　　D. 要除去這座高達4.5米的雕像上的塵土和油污

【答案及解析】

答案：C

如果考生答此題時，以「1-2-3」模式來進行閱讀分析，就會浪費時間。因為和答案相關的關鍵信息只為文段的首句（48字），從頭到尾閱讀完全段（133字）再看問題（20字），勢必還要回到文段中的第一句話（48字）尋求答案，這樣一來，來回的閱讀量是133+20+48，共191字。

但假如以「2-1-3」模式解題則不同，先看問題（20字）後，馬上反應出這是道「代詞指代」類題，那麼迅速在文中找到代詞的位置，進行閱讀分析第一句（48字）後，即可做出正確的判斷。那麼運用這樣一種模式，我們的閱讀量是20+48，合共只有68字。

當然，這一模式的優勢絕不是68字比191字這種單純的數量比較就能道明，但「語言理解與表達」確實應成為公務員考試中考生應拿分的部分，而且是「迅速」取分的部分。「2-1-3」模式是我們在應對語言理解類試題時必備的招式。

2. 聯想跳躍法

「聯想跳躍法」是快速閱讀的一種技巧，是指對於提供的材料不是每一個字都要「讀」，而是根據特定的「材料特點」或「提示性信息」，通過「聯想」達到的一種「跳躍式」閱讀的方法。

【例題】

1. 按照近代政治學理論觀點，「共和」的含義比較廣泛，涵蓋著民主概念。具體來說，共和國相對於君主國而言，凡是非君主制國家便是共和國，共和國在歷史上分成貴族共和國和民主共和國。應當說，這種政治理論觀點可能適用於近代國家情況，但不合乎古代政治觀念。共和國概念源於古羅馬，在西塞羅的著作中，共和國基本上相同於城市國家，用來表示一種國家形態。

    對這段文字主體的概括正確的一項是：
    A. 共和含義比較廣泛，涵蓋著民主概念
    B. 共和國的概念在古今有所不同
    C. 凡是非君主制國家便是共和國
    D. 共和國在歷史上分成貴族共和國和民主共和國

【答案及解析】

1. 答案：B

對比選項發現四個選項強調的重點有所不同。然後，看到材料一開頭就提到了「近代」——這是一個表示「時間、時期」的詞語，考生應該立即聯想到材料中有沒有「現代或者古代」？因為漢語表達的習慣，提到「時間」要麼是陳述事實、引入話題，要麼想表示「對比」。而在題目給定的材料中，所給的時間詞多數表示「前、後」情況的「對比」。所以在例題一中，看到「近代」這個詞，應該馬上聯想到材料中應該有「現代或者古代」，果然「聯想式跳躍」到下文中的表示時期的時間詞——「古代」，這時就很快判斷出材料的邏輯結構：由「近代」和「古代」兩個時間詞引導的一個「並列」的行文結構，答案也就很快選出來B。

【例題】

2. 孔子嘗曰：「未知生，焉知死？」生與死自孔子時起便是中國人始終關注的問題，並得到各種回答，尤其在漢代，人們以空前的熱情討論這兩個問題，不僅是出於學者的學術興趣，亦出於普通民眾生存的需要。然而，正如孔子所說，在中國思想史上，對生的問題的關注似乎遠勝於對死的問題的追問。有時候人們確實覺得後者更重要，但這並非由於死本身，而是因為人們最終分析認為，死是生的延續。

這段文字的核心觀點是：

A. 孔子關於生死的看法對中國人產生深遠影響
B. 生與死是中國思想史上長期受到關注的問題
C. 中國人對生與死的問題的討論實際以生為旨歸
D. 對生死問題的不同答案源自討論者的不同觀念

【答案及解析】

2. 答案：C

　　原文中的第一句話：「孔子嘗曰：未知生，焉知死？」這句話的意思是：「生的問題」還不清楚，怎麼研究「死」呢？意即對「生」的問題的關注、研究，是關注、研究「死」的問題的前提和基礎，所以材料的第一句話就點明了主題顯然是圍繞「生、死」並側重了對於「生」的問題的關注。

　　而在四個選項中，選項A、B、D三個選項中的「生、死」都是並列結構，即兩者都關注；而選項C中「生、死」中側重於「生」——與材料的第一句話吻合，答案直接選C即可。

　　即是說：300字左右的材料，大家只需要讀10個字，並弄懂「未知生，焉知死？」的意思即可；其他的內容都是在「聯想（後面的內容一定圍繞這句話展開論述）跳躍（大致瀏覽一下，意思沒有跑偏即可）閱讀」中——大概10秒鐘左右完成！

　　綜合上面所述，閱讀時一定要明確「提高閱讀速度」的目標，進而掌握、熟練而靈活地運用「聯想跳躍閱讀」的方法，在備考中不斷實踐、分析、總結、提升，相信大家一定會有驚喜的收穫。

3. 選擇答案的 4 個訣竅

1. 對數學運算比較有效的方法──聯繫法

聯繫法是指數字之間存在著一些必然聯繫，通過這些聯繫可以找出答案。比如在涉及距離速度的題中，出現了7和21、4和12等數字，你要聯想要答案可能跟3有關，而不是跟5、8等其他數字有關。

2. 對邏輯判斷比較有用的方法──驗證法

驗證法是指將選項帶入題幹的關鍵處來驗證其正確性的方法。邏輯判斷的題幹往往比較長，如果全部讀完，那是要花很多時間的，所以必須要簡化程序，直接將先帶到最後一句話的前後去檢驗，基本可以確定答案之所在。

3. 對語言理解與表達有效的方法──關鍵詞法

關鍵詞法是指對言語的理解要抓住重要的詞語，從而將其組織起來表達符合題幹的意思。在公務員考試中，語言理解與表達的題幹往往比較長，如果考生要認真地閱讀，有些題可能1分鐘都讀不完。這時候，考生就要進行速讀，將快速閱讀過後頭腦中殘存的信息組織起來，在答案中尋找具有相同形式或內容的選項。

4. 最簡單的辦法——造句法

　　造句法是指按照相關句式結構造出一個新句子的方法。造句法適用於類比推理和定義判斷。因為造句法的基本原理是相似事物之間具有異質同構性。

PART II VERBAL REASONING

Verbal Test Reasoning 就是給出根據一篇 100 至 200 多字的短文，判斷題幹信息正確與否，主要考察應聘者的英語閱讀能力和邏輯判斷能力。

2.1 True / False / Not Given 定義

　　Verbal Test Reasoning是給出根據一篇100至200多字的短文，然後讓考生判斷題幹信息正確與否，從中考察應聘者的英語閱讀能力和邏輯判斷能力。

答案有三個選擇：

　　1. True：題幹的信息根據原文來判斷，是正確的。

　　2. False：題幹的信息根據原文來判斷，是錯誤的。

　　3. Not Given（或Can't tell）：根據原文提供的信息，無法判斷對錯。

(A) True的定義：

　　1. 題目是原文的同義表達。通常用同義詞或同義結構。例如：

　　原文： Frogs are losing the ecological battle for survival, and biologists are at a loss to explain their demise（青蛙失去了生存下來的生態競爭能力，生物學家不能解釋它們的死亡。）

　　題目： Biologists are unable to explain why frogs are dying. （生物學家不能解釋為什麼青蛙死亡。）

解析：題目中的「are unable to」跟原文中的「are at a loss to」屬同義詞，題目中的「why frogs are dying」和原文中的「their demise」是同義詞，所以答案應為「True」。

2.題目是根據原文中的幾句話做出推斷或歸納。不推斷不行，但有時有些同學會走入另一個極端，即自行推理或過度推理。例如：

原文：Compare our admission inclusive fare and see how much you save. Cheapest is not the best and value for money is guaranteed. If you compare our bargain Daybreak fares, beware：most of our competitors do not offer an all inclusive fare.（比較我們包含的費用會看到你省了很多錢。最便宜的不是最好的。如果你比較我們的價格，會發現絕大多數的競爭對手不提供一籃子費用。）

題目：Daybreak fares are more expensive than most of their competitors.（Daybreak的費用比絕大多數的競爭對手都昂貴。）

解析：雖然文章沒有直接提到的費用比絕大多數的競爭對手都昂貴。但從原文幾句話中可以推斷出Daybreak和絕大多數的競爭對手相比，收費更高，但服務的項目要更全。與題目的意思一致，所以答案應為「True」。

(B) False的定義：

1. 題目與原文直接相反。通常用反義詞、not加同義詞及反義結構。no longer、not any more、not、by no means...對比used to do something、until recently、as was once the case。例如：

原文：A species becomes extinct when the last individual dies.（當最後一個個體死亡時，一個物種就滅亡了。）

題目：A species is said to be extinct when only one individual exists.（當只有一個個體存活時，一個物種就被說是滅亡了。）

解析：可以看出題目與原文是反義結構。原文說一個物種死光光，才叫滅絕，而題目說還有一個個體存活，就叫滅絕，題目與原文直接相反，所以答案為「False」。

2. 原文是多個條件並列，題目是其中一個條件（出現must、only）原文是兩個或多個情形（通常是兩種情形）都可以。常有both...and、and、or及also等詞，及various、varied、variety、different、diversified、versatile等表示多樣性的詞彙。題目是「必須」或「只有」或是「單一」其中一個情況，常有must、only、 sole、one、single等詞。例如：

原文：Booking in advance is strongly recommended as all Daybreak tours are subject to demand. Subject to availability, stand by tickets can be purchased from the driver.（提前預定是強烈建議的，因為所有的Daybreak旅行都是由需求決定的。如果還有票的話，可直接向司機購買。）

題目： Tickets must be bought in advance from an authorized Daybreak agent.（票必須提前從一個認證的代理處購買。）

解析： 原文是提前預定、直接向司機購買都可以，是多個條件的並列。題目是必須提前預定，是必須其中一個情況。所以答案應為「False」。

3. 原文強調是一種：理論、感覺、傾向性、期望或是預測等詞, 而題目強調是一種「事實」，例如：

原文： But generally winter sports were felt to be too specialized.（但一般來説，冬季項目被感覺是很專門化的。）

題目： The Antwerp Games proved that winter sports were too specialized.（Antwerp運動會證明冬季項目是很專門化的。）

解析： 原文中有feel，強調是「感覺」。題目中有prove，強調是「事實」。所以答案應為「False」。

4.原文和題目中使用了表示不同程度、範圍、頻率、可能性的詞。原文中常用typical、odds、many（很多）、sometimes（有時）及unlikely（不太可能）等詞。題目中常用special、all（

全部）、usually（通常）、always（總是）及impossible（完全
不可能）等詞。例如：

原文：Frogs are sometimes poisonous.（青蛙有時是有毒的）

題目：Frogs are usually poisonous.（青蛙通常是有毒的）

解析：原文中有sometimes，強調是「有時」。題目中有
usually，強調是「通常」。所以答案應為「False」。

5. 情況原文中包含條件狀語，題目中去掉條件成份。原文中
包含條件狀語，如if、unless或if not也可能是用介詞短語表示條件
狀語如in、with、but for或exept for。題目中去掉了這些表示條件
狀語的成份。這時，答案應為False。例如：

原文：The Internet has often been criticized by the media as
a hazardous tool in the hands of young computer users.（Inter-
net通常被媒體指責為是年輕的電腦用戶手中的危險工具。）

題目：The media has often criticized the Internet because it
is dangerous.（媒體經常指責Internet ，因為它是危險的。）

解析：原文中有表示條件狀語的介詞短語 in the hands of young
computer users, 題目將其去掉了。所以答案應為「False」。

6. 出現以下詞彙，題目中卻沒有說明：less obviously、less
likely或less possible。

(C) Not Given的定義

1. 題目中的某些內容在原文中沒有提及。題目中的某些內容在原文中找不到依據。

2. 題目中涉及的範圍小於原文涉及的範圍，也就是更具體。原文涉及一個較大範圍的範疇，而題目是一個具體概念。也就是說，題目中涉及的範圍比原文要小。例如：

原文：Our computer club provides printer.（我們電腦學會提供打印機）

題目：Our computer club provides color printer.（我們電腦學會提供彩色打印機）

解析：題目中涉及的概念「比原文中涉及的概念」要小。換句話說，電腦學會提供打印機，但是究竟是彩色還是黑白的，不知道或有可能，文章中沒有給出進一步的信息。所以答案應為「Not Given」。

3. 原文是某人的目標、目的、想法、願望、保證、發誓等，題目是事實。原文中常用aim、goal、promise、swear、vow、pledge、oath、resolve等詞。題目中用實意動詞。例如：

原文：He vowed he would never come back.（他發誓他將永不回來。）

題目： He never came back. （他沒再回來。）

解析： 原文中説「他發誓將永不回來」，但實際怎麼樣，不知道。也可能他違背了自己的誓言。所以答案應為「Not Given」。

4. 原文中沒有比較級，題目中有比較級。例如：

原文： In Sydney, a vast array of ethnic and local restaurants can be found to suit all palates and pockets. （在悉尼，有各種各樣的餐館。）

題目： There is now a greater variety of restaurants to choose from in Sydney than in the past. （現在有更多種類的餐館可供選擇）

解析： 原文中提到了悉尼有各種各樣的餐館，但並沒有與過去相比，所以答案應為「Not Given」。

5. 原文中是虛擬「would / even if」，但題目中卻是事實。
（虛擬語氣看到當作沒有看到）

6. 原文中是具體的數據事例，而題目中卻把它擴大化，規律化。

2.2 True / False / Not Given 答題技巧

　　對於「是/非/無判斷題」（True/False/Not Given），讓大多數考生十分困惑的就是False和Not Given之間的區別。

　　以下試看一例：

原文： Sydney harbour is one of the largest in the world.

題幹： Sydney harbour is the largest in the world.

　　有同學就說：我知道，世界上最大的港口是荷蘭的鹿特丹，而非悉尼港，因此這道題是False，但Verbal Reasoning的判斷題是語言類的考試，不是常識考試，判斷一道題到底是True、False，還是Not Given，就要從這三點基本原則出發，即原文和題幹之間是一致性則為True，矛盾性則為False，不確定性則為Not Given。比如上文的原文說悉尼港是世界第一大港口之一，那麼有可能悉尼港是第一大，也有可能是第二大，因此題幹的表述「悉尼港是世界第一大港口」就是不確定的，因此為Not Given。

關於判斷題的讀題技巧，本文主要介紹以下六點原則，主要幫助考生在讀題時定位考點，破解題目：

1. 信號詞原則：先在題幹中劃出定位信號詞，要注意名詞是關鍵，但要挑選「長相別致」的名詞，比如：時間、數字、年代、人名、地名、機構名稱、專有名詞、特殊稱謂、大寫字母、斜體字等。這些信號詞有助於考生回原文快速而準確定位題幹出處。

2. 假如題幹沒有明顯信號詞的話，則可根據判斷題的出題特點採用關係原則和順序原則：

a. 出題特點
從出題特點來看，判斷題的考點除了比較明確的要考數字時間範圍的精確度以外，更多的是考兩個事物之間的「關係」，包括並列、因果、遞進、轉折、主被動等明顯關係，以及隱藏著的隱性「關係」，因此審題時先找出題幹中產生關係的兩個事物A和B，然後以A和B為定位關鍵詞回原文定位，具體情況見下：

①A和B都在文中出現，但文中兩者並無任何關係，即題幹中的關係在文中不存在，此時答案即為Not Given；

②A和B存在，但文中關係表述與題幹相抵觸，則答案為False；

③A和B存在，文中表述的兩者關係和題幹中的為同義轉換，即表述一致，那麼答案為True；

④A和B中有一者在文中未出現，且沒有出現他們的近義或反義表達，則答案也為Not Given，因為產生「關係」的事物都沒有全部出現，就更不用説兩者的關係了。

b. 順序原則

第二，從順序原則的角度來看，判斷題的出題順序基本與原文行文一致，因此要先解容易定位或者相對容易判斷題幹中關系的題目，縮小答案的搜索範圍後再解其餘題目。例如，第一和第二小題沒有明確定位關鍵詞，而第三小題題幹中有大寫的地名或年代容易回原文定位，則可先對第三小題進行定位，然後在第三題的文中相對應部分之前的文章內容中尋讀第一和第二小題的對應點。

以下是另外幾道題目：

1. Early peoples found it easier to count by using their fingers than a group of pebbles.

2. For the earliest tribes, the concept of sufficiency was more important than the concept of quantity.

3. Indigenous Tasmanians used only four terms to indicate numbers of objects.

4. Some peoples with simple number systems use body language to prevent misunderstanding of expressions of number.

5. All cultures have been able to express large numbers clearly.

6. The word 'thousand' has Anglo-Saxon origins.

7. In general, people in seventh-century Europe had poor counting ability.

8. In the Tsimshian language, the number for long objects and canoes is expressed with the same word.

9. The Tsimshian language contains both older and newer systems of counting.

解題思路：

應先解3、6、7、8及9這些題中有大寫字母或年代這些相對容易定位的題目，例如33題在文中對應的部分為第三小節開頭，則根據順序原則第2題在文中的對應部分最後可能出現在開頭兩個小節中，這樣一來尋讀2題的範圍就縮小了。

再者，以「關係」的視角來解題，那麼第2題可以被看成是the concept of sufficiency 與the concept of quantity之間的比較關係，以這兩項回原文尋讀，對應點為「Our ancestors had little use for actual numbers; instead their considerations would have been more of the kind. Is this enough? Rather than how many when they were engaged in food gathering, for example.」

此句以「is this enough」和「How many」兩個短句來同義替代題幹中的兩個concept，而句中提到的我們的祖先考慮較多的是夠不夠而不是有多少則來同義轉換題幹中的more important，因此由於比項均出現且兩者的「關係」在文中也有同義表達，此題答案即為True。

此外，第1題也可以看作是有「關係」的題目，即fingers與groups of pebbles之間的關係，用哪一個數數更容易，文中對應點位「Counting is not directly related to the formation of a number concept because it is possible to count by matching the items being counted against a group of pebbles, grains of corn, or the counter's fingers.」雖然產生「關係」的手指與鵝卵石均有重現，但兩者並無任何比較關係，因此答案為Not Given。

3. 注意否定考點詞：否定考點詞是判斷題中非常高頻、重要及優先的考點詞，除了not、never、hardly、few、little，以及某些否定前綴如dis-、mis-、in-、un-等之外，考生還需要了解一些隱含或間接的否定詞，如：fail to、be at a loss to、be free of、

lack、be short of等等。例如：

題幹：London Zoo's advertisements are dishonest.

原文：One of London Zoo's recent advertisements caused me some irritation, so patently did it distort the reality.

首先確定題目中的考點為否定詞dishonest，否定考點詞在思路上一定要回原文找否定對應，具體做法是將考點詞dishonest一分為二，分成「dis-」與「honest」兩部分，我們現在只需回原文找到這兩個部分，其分別的對應點就可以確定答案為True了，如果發現方向是相反的，則False，沒有講是否誠實的話，那就選Not Given。結果在原文中發現了distort中的否定前綴dis-，而在其後又找到了honest的對應詞彙reality，答案直接鎖定True。即使不認識distort這個詞，或者說原文中的這個倒裝句根本看不懂，但是只要你掌握了這種對應法，一切問題都可迎刃而解。

4. 考點唯一原則：有的考生還會有個疑問，原文中講的是倫敦動物園的廣告之一，而題目中則泛指了倫敦動物園的廣告，為什麼不是Not Given，因為不清楚是不是所有的廣告都不誠實。做這種題要採用考點唯一原則，即每道題目只設置一個陷阱，大家又問：那這個唯一的考點為什麼一定是dishonest，而不是adver-tisements？

　　道理很簡單，否定詞是優先考點。那麼除了否定詞是考點外，還有哪些必然屬於考點呢？題幹中必是考點的包括：數字對應的內容（時間、增降幅度）、比較級、最高級、絕對詞、否定詞、句內關係。

　　再看一例：

題幹：Zoos made an insignificant contribution to conservation up until 30 years ago.

原文：Zoos were originally created as places of entertainment, and their suggested involvement with conservation didn't seriously arise until about 30 years ago.

　　這道題目的考點換成了insignificant，這次的對應詞不像上一道題目那麼好找，in這次對應是didn't，而significant對應的是seriously arise，答案同樣為True。

　　還是這篇文章裡的一道題目：

題幹：The number of successful zoo conservation programs is unsatisfactory.

原文：Today approximately 16 species might be said to

have been 'saved' by captive breeding programs, although a number of these can hardly be looked upon as resounding successes.

現在大家一定已經清楚了考點詞為unsatisfactory，再回原文找一下對應，這次un-對應hardly，而satisfactory對應的是success。

綜合兩道題目來看，考點都為否定詞，做題目的思路上基本一致，都在找對應，而答案又驚人地一致，都選了True，請仔細體會。

下面再給大家一道題目，練習一下：

題幹：Biologists are unable to explain why the frogs are dying.

原文：Biologists are at a loss to explain their demise.

能不能找到unable的對應詞呢？如果你並不認識原文中的間接否定短語be at a loss to的話，那麼至少你知道loss這個詞能夠表明否定含義，否定詞對應到了否定詞，其他的就不用擔心了。答案為True。原文中的短語be at a loss to 意思即為「不知道，不知所措。」請注意積累。

5. 中心詞原則：如果題幹中的名詞是文章中心詞，比如在文章的標題中出現，那麼這個名詞絕不是信號詞，因為很多段落中都會出現該名詞，很難判斷出題目落在哪個段落。

6. 轉換詞性原則：若在題幹中無法找到合適的名詞，可以換其他詞性→形容詞和副詞的比較級、最高級和反義詞；動詞的同義詞、反義詞和該動詞的範圍程度；數詞在英語和阿拉伯數字的互相轉換（50/Half）；連詞在文章本身中隱含答案，最常暗示答案的幾個連詞是But、However、While、And、Also、Moreover，考試對冠詞、介詞、代詞、感嘆詞這四種詞性基本不進行考查。

2.3 Verbal Reasoning Test: Question

Passage 1

Instituted in 1979 as a temporary measure to limit population growth, China's one child policy remains in force today and is likely to continue for another decade. China's population control policy has attracted criticism because of the manner in which it is enforced, and also because of its social repercussions. Supporters of the Chinese government's policy consider it a necessary measure to curb extreme overpopulation, which has resulted in a reduction of an estimated 300 million people in its first twenty years. Not only is a reduced population environmentally beneficial, it also increases China's per capita gross domestic product. The one-child policy has led to a disparate ratio of males to females – with abortion, abandonment and infanticide of female infants resulting from a cultural preference for sons. Furthermore, Draconian measures such as forced sterilization are strongly opposed by critics as a violation of human reproduction rights. The one-child policy is enforced strictly in urban areas, whereas in provincial regions fines are imposed on families with more than one child. There are also exceptions to the rules – for example, ethnic minorities. A rule also allows couples without siblings to have two children – a provision which applies to millions of sibling-free adults now of child- bearing age.

Q1. China's one-child policy increases the country's wealth.

True   False   Not Given

Q2. The passage suggests that two-child families will dramatically increase, as sibling-free adults reach child-bearing age.

True   False   Not Given

Q3. The main criticism of China's one-child policy is that it violates human rights.

True   False   Not Given

Q4. Families with more than one child are more common in China's rural areas.

True   False   Not Given

Q5. The general preference among Chinese parents is for male babies.

True   False   Not Given

Passage 2

There are 562 federally recognized American Indian tribes, with a total of 1.7 million members. Additionally, there are hundreds of groups seeking federal recognition – or sovereignty – though less than ten percent will successfully achieve this status. Federally recognised tribes have the right to self-government, and are also eligible for federal assistance programmes. Exempt from state and local jurisdiction, tribes may enforce their own laws, request tax breaks and control regulatory activities. There are however limitations to their sovereignty including, amongst others, the ability to make war and create currency. Historically, tribes were granted federal recognition through treaties or by executive order. Since 1978 however, this has been replaced by a lengthy and stringent regulatory process which requires tribes applying for federal recognition to fulfil seven criteria, such as anthropological and historical evidence. One of the complications regarding federal recognition is the legal definition of "Indian". Previously, racial criteria, tribal records and personal affidavits were used to classify American Indians. Since the 1970s, however, there has been a shift to the use of a political definition – requiring membership in a federally recognized tribe in order to qualify for benefits, such as loans and educational grants. This definition, however, excludes many individuals of Native American heritage who are not tribal members.

Q6. There are only two exemptions to a federally recognized tribe's powers of self- government.

True   False   Not Given

Q7. Demand for federal recognition is high because it is a prerequisite for benefit programmes.

True   False   Not Given

Q8. Since 1978 it has become harder for a tribe to achieve federally recognized status.

True   False   Not Given

Q9. Federally recognized tribes are not subject to state laws and do not pay taxes.

True   False   Not Given

Q10. A large number of people who identify themselves as American Indians do not fulfill the legal definition.

True   False   Not Given

Passage 3

Sodium chloride, or salt, is essential for human life. Typically derived from the evaporation of sea water or the mining of rock salt deposits, salt has been used by humans for thousands of years as a food seasoning and preservative. The mineral sodium is an electrolyte – an electrically-charged ion – that enables cells to carry electrical impulses to other cells, for example muscle contractions. Electrolytes also regulate the body's fluid levels. A diet deficient in salt can cause muscle cramps, neurological problems and even death. Conversely, a diet high in salt leads to an increased risk of conditions such as hypertension, heart disease and stroke. In spite of high-profile campaigns to raise awareness, salt consumption has increased by 50% in the past four decades, with the average adult ingesting more than double the amount of salt their body requires. Much of this increase can be attributed to the advent of frozen and processed foods in the mid-twentieth century. In the United States it is estimated that excessive salt consumption claims 150,000 lives and results in $24 billion of health care costs annually. For individuals wishing to reduce their sodium intake, the answer is not simply rejecting the salt shaker; 75% of the average person's salt consumption comes from food, such as bread, cereals, and cheese.

Q11. Humans primarily use salt for food flavouring and preservation.

True   False   Not Given

Q12. Most adults consume 50% more salt than their body requires.

True   False   Not Given

Q13. Frozen and processed foods contain no more salt than contained in a typical diet.

True   False   Not Given

Q14. Over three quarters of the average person's salt consumption comes from frozen foods.

True   False   Not Given

Q15. The human body needs salt to maintain constant levels of body fluids.

True   False   Not Given

Passage 4

The United Nations' Convention on International Trade in Endangered Species (CITES) recently reaffirmed a 1989 ban on trading ivory, despite calls from Tanzania and Zambia to lift it. Only 470,000 elephants remain in Africa today – compared to 1.3 million in 1979. While natural habitat loss was a significant factor in dwindling elephant populations, poaching for ivory was the main cause. Since the ban's implementation, elephant populations have recovered in many African countries, but an estimated38,000 elephants are still killed annually. CITES permitted one-off sales in 1999 and in 2008, allowing approved countries to dispose of their government stockpiles of ivory. Ivory from these sales was exported to Japan and China, where demand for carved ivory is high. Conservation groups vehemently oppose further one- off sales, because much of the ivory sold is of unknown origin. Furthermore, the sales have fuelled far-Eastern demand for ivory. In central and western African countries, where organized crime rings operate lucrative ivory smuggling operations, poaching remains rife. Those in favour of allowing one-off sales argue that elephants are no longer endangered, and that maintaining the ban will simply inflate the price of illegal ivory, making poaching more tempting. Though the CITES decision is viewed as a victory by conservationists, the African elephant's future relies on governments' commitment to enforcing the ban.

Q16. No legal sales of Ivory have occured since 1989. True False

    True   False   Not Given

Q17. Whether or not African elephants should be classified as endangered is debatable.

    True   False   Not Given

Q18. Conservationists question the provenance of the ivory sold at one-off sales.

    True   False   Not Given

Q19. Because their elephant populations are thriving, Tanzania and Zambia want to lift the ban on ivory trading.

    True   False   Not Given

Q20. Increased demand from Japan and China is driving up the price of ivory.

    True   False   Not Given

Passage 5

In biology, the term mutualism refers to a mutually beneficial relationship between two species. The later economic theory of mutualism is based on the labour theory of value, which states that the true cost of an item is the amount of labour that was required to produce it. Hence, goods should not be sold for more than the cost of acquiring them. Mutualism is closely associated with anarchism, because its principles were set forth in the mid- nineteenth century by the French politician and philosopher Pierre-Joseph Proudhon – the first person to define himself as an "anarchist". The main tenets of mutualism are free association and free credit. In a mutualist workplace, workers with different skills form an association to create a product or service. Though dependent on each other, the workers are not subordinated as in a capitalist enterprise. Mutual banks, also called credit unions, operate on the belief that free credit enables profit to be generated for the benefit of the union's members rather than bankers. Modern-day mutualism is sometimes described as free-market socialism. Proponents of mutualism support a free market economy, but object to capitalism because of the inequalities created by government intervention. Many mutual businesses and banking establishments exist today, espousing Proudhon's Co-operative model.

Q21. Proudhon's economic theory of mutualism was influenced by biological mutualism.

True   False   Not Given

Q22. Mutual banking establishments do not operate on a
for-profit basis. True   False   Not Given

True   False   Not Given

Q23. The labour theory of value is defined as: only the
person who made an item should profit from its
sale.

True   False   Not Given

Q24. In common with socialism, the economic doctrine
of mutualism advocates state control over production.

True   False   Not Given

Q25. Free association separates labour from hierarchy
and ownership.

True   False   Not Given

Passage 6

Stem cells are cells that can self-renew and differentiate into specialised cell types. Because of their potential to replace diseased or defective human tissue, stem cells are seen by scientists as key to developing new therapies for a wide range of conditions, including degenerative illnesses and genetic diseases. Treatments based on adult stem cells – from sources such as umbilical cord blood – have been successfully developed, but what makes stem cell research controversial is the use of embryonic stem cells. Not only do embryonic stem cells reproduce more quickly than adult stem cells, they also have wider differentiation potential. The main opponents to stem cell research are pro-life supporters, who believe that human life should not be destroyed for science. Advocates of stem cell research counter this crucial point by saying that an embryo cannot be viewed as a human life, and that medical advances justify the destruction of embryos. Furthermore, stem cell research utilises the thousands of surplus embryos created for in vitro fertilisation treatment. The issue is particularly divisive in the United States, where federal funding is not available for the creation of new embryonic stem cell lines, although recent legislation has opened up government funding to further research on embryonic stem cells created through private funding. Whereas many governments prohibit the production of embryonic stem cells, it is allowed in countries including the UK, Sweden and the Netherlands.

Q26. Stem cells are at the forefront of medical research because of their ability to grow indefinitely.

True   False   Not Given

Q27. The United States government does not supply funding for projects using embryonic stem cell lines.

True   False   Not Given

Q28. One advantage of embryonic stem cells over adult stem cells is their greater ability to be converted into specialised cell types.

True   False   Not Given

Q29. The bioethical debate over embryonic stem cell research centres on whether it involves the destruction of human life.

True   False   Not Given

Q30. Treatments based on embryonic stem cells provide therapies for a wide range of diseases.

True   False   Not Given

Passage 7

The project was ambitious in its size, complexity, triparty nature, and in its pioneering of the Private Finance Initiative. This difficulty was unavoidable and contributed to the project's failure. However, a more thorough estimate of the unknown difficulties and timescales would have enabled the Department to better prepare for the project, and increase its chance of success.

In December 1997 XSoft indicated they needed more time to complete the project, which should have been inevitable. If the Department knew from the start how long the project would take, it is questionable whether they would have considered inception, especially considering the implications of delay on the overall profitability for the venture.

Q31. If more care had been put into estimating the difficulties, it is less likely the project would have failed.

True   False   Not Given

Q32. XSoft witheld information from the Department regarding how long the project would take.

True   False   Not Given

Q33. The Department's profits were dependent upon how long the project took.

True   False   Not Given

Passage 8

Ever since the gun's invention it has been changing the world in many different ways. Many of the developments in gun design have been brought about by man's desire to protect himself, and the challenge of inventing bigger and more accurate weapons.

Each time there has been a major innovation in the development of the gun, there has been a profound effect on the world. The gun helped in the exploration of the world, it has also helped in the development of society as we know it.

Q34. The gun was invented because the human race needs to protect themselves.

True   False   Not Given

Q35. Guns are the reason our society is the way it is today.

True   False   Not Given

Q36. Financial incentives had no part to play in the development of the gun.

True   False   Not Given

Passage 9

Being socially responsible is acting ethically and showing integrity. It directly affects our quality of life through such issues as human rights, working conditions, the environment, and corruption. It has traditionally been the sole responsibility of governments to police unethical behaviour. However, the public have realised the influence of corporations and, over the last ten years, the level of voluntary corporate social responsibility initiatives that dictate the actions of corporations has increased.

Q37. The ethical actions of corporations has changed over the last ten years.

  True   False   Not Given

Q38. Corporations can influence the public's quality of life. True False

  True   False   Not Given

Q39. Traditionally the government have relied upon only the large corporations to help drive corporate social responsibility, whilst they concentrated on the smaller corporations.

  True   False   Not Given

Passage 10

A well-nourished child can be more likely to be a studious one. But food has been seen as a cost to be cut in times of austerity, rather than an ingredient of good schooling. That may now be changing: as the government worries about obesity – which is fast rising among children- and urges everyone to eat less salt, fat and sugar, and more fruit and vegetables, the deficiency and unhealthiness of most school meals is striking. But cash constraints make change difficult.

Q40. Children who eat healthily will perform better in exams.

   True　　　False　　　Not Given

Q41. The level of child obesity has differed over the years. True False

   True　　　False　　　Not Given

Q42. The government is apathetic about obesity. True False

   True　　　False　　　Not Given

2.4 Verbal Reasoning Test: Answer

Q1. China's one-child policy increases the country's wealth.

Not Given - The fourth sentence states that the policy increases China's per capita gross domestic product, however this is just one economic indicator and is based on output per person. The passage does not tell us if overall, the country as a whole has increased wealth due to the one child policy. Since the passage does not tell us either way, we must respond Cannot Say.

Q2. The passage suggests that two-child families will dramatically increase, as sibling-free adults reach child-bearing age.

False - The last sentence merely presents the fact that millions or sibling-free couples are able to have two children, and does not speculate as to the implications.

Q3. The main criticism of China's one-child policy is that it violates human rights.

Not Given - The second sentence states that both the policy's manner of enforcement and its social repercussions are criticised – but does not state which is the main criticism. So based on the information we are given, we cannot say.

Q4. Families with more than one child are more common in China's rural areas.

Not Given - While the 7th sentence states that the policy is enforced less strictly in provincial regions, comparative figures are not pro vided. One might deduce this statement is likely given what we are told, but we are not told explicitly if this is true or false, therefore we cannot say.

Q5. The general preference among Chinese parents is for male babies.

True - The fifth sentence tells us that "a disparate ratio of males to females" is the result of "a cultural preference for sons". Whilst it might be impossible to make assumptions about what each parent's preference is, the key word in the statement in "general" which means we can look at the overall trend, in this case towards sons.

Q6. There are only two exemptions to a federally recognized tribe's powers of self- government.

False - The fifth sentence states that there are "limitations" and cites two "amongst others". So we are told there are more than two limitations.

Q7. Demand for federal recognition is high because it is a prerequisite for benefit programmes.

Not Given - The passage does not specify why hundreds of groups are seeking federal recognition. Even though one might postulate demand is due to receiving benefits, the passage does not tell us this is or is not the case, so we cannot say.

Q8. Since 1978 it has become harder for a tribe to achieve federally recognized status.

Not Given – The 7th sentence refers to the introduction of a "lengthy and stringent regulatory process" in 1978 however the passage does not tell us how difficult it was before the introduction of this process. In order to respond either True or False we would need to know about before and after 1978 to draw a comparison, but we do not so we must respond Cannot say.

Q9. Federally recognized tribes are not subject to state laws and do not pay taxes.

False – While the fourth sentence states that tribes are "exempt from state and local jurisdiction" but goes on to say that the tribes may "request tax breaks". So as we are told tribes may request state tax breaks, this tells us they must normally pay taxes.

Q10. A large number of people who identify themselves as American Indians do not fulfill the legal definition.

True – The second sentence states that there are hundreds of groups attempting to attain federal recognition. The final two sentences states that the legal definition of Indian is membership in a federally recognised tribe. Thus, it follows that many American Indians do not fulfil the legal definition.

Q11. Humans primarily use salt for food flavouring and preservation.

Not Given – Though these are, indeed, two uses of salt as stated in the second sentence, we are not told whether these uses are the primary use of salt. As an aside, food use actually accounts for less than 20% of salt production.

Q12. Most adults consume 50% more salt than their body requires.

Not Given – The passage tells us that "the average adult" ingests "more than double the amount of salt their body requires". However we are not told how this average is distributed in order to say whether this statement is true or not. For example it could be possible that the average salt intake is skewed by a small proportion of adults. We cannot tell from the information given alone.

Q13. Frozen and processed foods contain no more salt than contained in a typical diet.

False – The 7th sentence states that salt consumption has increased 50%, and the
8th sentence states "much of this increase can be attributed to the advent of frozen and processed foods". So in order for salt consumption to increase, the salt levels in the new food (frozen and processed food) must be above the average level, or disproportionately high.

Q14. Over three quarters of the average person's salt consumption comes from frozen foods.

False – The last sentence of the passage states that 75% of the average person's salt intake comes from "food, such as bread, cereals and cheese". Even if all of this food were classed as frozen (highly implausible) this would still only get us to 75% of the average person's salt intake; it would not get us to 'over three quarters' as the statement says. So it cannot be true.

Q15. The human body needs salt to maintain constant levels of body fluids.

Not Given – The third and fourth sentences tell us that salt contains electrolytes, and that electrolytes "regulate the body's fluid levels". However we are not told if these electrolytes can be found in other foods or drinks, thus we cannot say.

Also, don't be misled by the first sentence "salt, is essential for human life"; since this sentence alone does not tell us that salt is essential to maintain levels of body fluids.

Q16. No legal sales of Ivory have occurred since 1989.

False – The passage states that two permitted one-off Ivory sales occurred in 1999 and 2008.

Q17. Whether or not African elephants should be classified as endangered is debatable.

True – The passage presents facts about the on-going problem of poaching, but also states that populations have recovered in many countries and that proponents of one-off sales "argue that elephants are no longer endangered". Based on the discussion, and the fact that we are told those in favour "argue" that elephants are no longer endangered, we can say the matter is debateable.

Q18. Conservationists question the provenance of the ivory sold at one-off sales.

True – The seventh sentence states that conservation groups oppose further sales because "much of the ivory sold is of unknown origin".

Q19. Because their elephant populations are thriving, Tanzania and Zambia want to lift the ban on ivory trading.

Not Given – Though Tanzania and Zambia want to lift the ban, it is not specified in the passage that they are two countries in which elephant populations have recovered – or indeed whether this is the reason for their calls to lift the ban.

Q20. Increased demand from Japan and China is driving up the price of ivory.

Not Given – The sixth and eight sentences state that there is demand for ivory from China and Japan, the tenth sentence suggests it is the ban – rather than the demand–that is inflating prices. Either way, the passage does not unequivocally state this is or is not the case, therefore we cannot say.

Q21. Proudhon's economic theory of mutualism was influenced by biological mutualism.

Not Given – The passage does not state whether the economic theory was influenced by biology, although it did come later.

Q22. Mutual banking establishments do not operate on a for-profit basis.
True
False – Sentence eight tells us that mutual banking establishments do seek to "generate money", but this profit is shared between union members rather than the bankers. So regardless of who it goes to, we are told the banks do create profit.

Q23. The labour theory of value is defined as: only the person who made an item should profit from its sale.

False – This is not the definition provided in the second sentence. We are told that the labour theory of value is that the true cost of an item is the amount of labour required to produce it.

Q24. In common with socialism, the economic doctrine of mutualism advocates state control over production.

False – The passage explains that free association is when workers form an association to create a product, thus production is controlled by the workers rather than the state. And while the ninth sentence refers to mutualism as free-market socialism, the tenth sentence states that proponents of mutualism object to the "inequalities created by government intervention".

Q25. Free association separates labour from hierarchy and ownership.

Not Given – Although we are told that in free association, "workers are not subordinated as in a capitalist enterprise", the passage does not directly discuss ownership.

In-fact in mutualistic workplaces, the organisation is often owned by the labour force themselves and/or the organisations customers/ stakeholders.

Q26. Stem cells are at the forefront of medical research because of their ability to grow indefinitely.

Not Given – The passage does not tell us if stem cells can grow "indefinitely". Also, the second sentence tells us that stem cells are seen as "key to developing new therapies". It would be a stretch to interpret this as meaning they are at the forefront of medical research, especially as we are not told about their significance compared with other areas of research.

Q27. The United States government does not supply funding for projects using embryonic stem cell lines.

False – The 8th sentence states that federal funding is now available for further research into stem cells lines that have been created using private funding.

Q28. One advantage of embryonic stem cells over adult stem cells is their greater ability to be converted into specialised cell types.

True – The first sentence defines cell differentiation as changing into specialised cell types. The fourth sentence states that embryonic stem cells have a "wider differentiation potential" than adult stem cells. Therefore the statement is true.

Q29. The bioethical debate over embryonic stem cell research centres on whether it involves the destruction of human life.

True – The fifth and sixth sentences present either side of this debate. The debate can be said to 'centre' on this issue because the passage tells us "the main opponents to stem cell research are pro-life supporters" and then "Advocates of stem cell research counter this crucial point".

Q30. Treatments based on embryonic stem cells provide therapies for a wide range of diseases.

Not Given – The third sentence states that treatments based on adult stem cells have been developed. However it does not specify whether treatments based on embryonic stem cells have also been successfully developed.

Q31. If more care had been put into estimating the difficulties, it is less likely the project would have failed.

True - Since more thorough can be considered equivalent to giving more care, this is stated as true in the following excerpt "a more thorough estimation of the unknown difficulties and timescales would increase its chance of success".

Q32. XSoft witheld information from the Department regarding how long the project would take.

Not Given - We are told XSoft requested more time, and the passage implies the Department did not know how long it would take at the beginning, but the passage does not tell us if XSoft did or did not withhold time information.

Q33. The Department's profits were dependent upon how long the project took.

True - We are told that there were "implications of delay on overall profitability for the venture".

Q34. The gun was invented because the human race needs to protect themselves.

Not Given - The passage does not say how or why the gun was invented. It does say that some changes to the gun's design have been because humans want to protect themselves, but the passage does not say how or why the gun was first invented.

Q35. Guns are the reason our society is the way it is today.

False - The gun "has also helped in the development of society as we know it". The word help implies it is not the only contributor and is therefore not the reason our society is the way it is today.

Q36. Financial incentives had no part to play in the development of the gun.

Not Given - The passage does not mention financial incentives or economic gain, so we cannot tell from information in the passage alone.

Q37. The ethical actions of corporations has changed over the last ten years.

True - The passage says that the public have caused corporations to alter their ethical actions: "over the last ten years, the level of voluntary corporate social responsibility initiatives that dictate the actions of corporations has increased."

Q38. Corporations can influence the public's quality of life.

True - This one is slightly less obvious. We are first told that being socially responsible directly affects our quality of life. Then we are told that corporate social responsibility dictates the actions of corporations. So following that logic, corporations must be able to affect our quality of life.

Q39. Traditionally the government have relied upon only the large corporations to help drive corporate social responsibility, whilst they concentrated on the smaller corporations.

False - We are told that "traditionally it has been the sole responsibility of the government to police unethical behaviour", meaning traditionally no corporation played a part; not even large corporations.

Q40. Children who eat healthily will perform better in exams.

Not Given- The passages says "a well-nourished child can be more likely to be a studios one", which means they are more likely to study. Whilst the passage suggests eating healthily can help, it does not say definitively if the child will or will not perform better in exams.

Q41. The level of child obesity has differed over the years.

True - The passage says, in reference to obesity, "is fast rising among children" meaning it has changed over time.

Q42. The government is apathetic about obesity.

False - The passage says " as government worries about obesity ". Worrying about something is the opposite of being apathetic about something.

PART III DATA SUFFICIENCY TEST

Data Sufficiency Test 不但要求考生具備基礎的數學知識和熟練的計算技巧，而且注重檢驗考生分析定量問題的能力，即根據已給出的數據，辨認哪些數據與問題有關，確定在何種情況下所給數據能滿足問題的要求，以此來檢驗考生的推理和綜合分析的能力。

3.1 Data Sufficiency Test 答題攻略

　　數據填充題（Data Sufficiency）不僅要求考生具備基礎的數學知識和熟練的計算技巧，而且十分注重檢驗考生分析定量問題的能力。接下來為大家講解一下注意事項，希望對大家備考有幫助。

　　數據填充題需要考生根據已給出的數據，辨認哪些數據與問題有關，確定在何種情況下所給數據能滿足問題的要求，以此來檢驗考生的推理和綜合分析的能力。在做數據填充題時，考生應注意以下幾點：

　　1. 不用求出精確的數字答案，只要根據已給的數據找到答案。

　　2. 即使發現數據（1）足以答題，也千萬不要倉促地選擇A，而應該繼續審題，看數據是否也能單獨解題。如果數據（2）也能解題，則應選擇D。

　　3. 應試人應當熟悉某些必需的日常生活知識。例如某題提到閏年，我們就應該想到，閏年的二月份只有28天，而且應將這一數據考慮到原題中，不要因為數據（1）和（2）沒有提到它而將其忽略了。

4. 涉及到幾何圖形時，千萬不要依賴試卷上給出的圖形而得出錯誤的假設和判斷。有時從圖形上看似乎並非全是按比例繪制的。

不少考生往往在這部分的得分比較低，一是由於題型不熟悉；二是答題速度太慢。為了提高考生的應試能力，我們推薦一種較為合理的解題方法，供大家平時練習和考試中使用。

在選擇答案前，首先回答下列三個問題：

1. 第一個說明能否單獨求解問題？

2. 第二個說明能否單獨求解問題？

3. 兩個說明放在一起能否求解問題？

如果問題（1）的答案是肯定的，那麼可能的選擇答案是A或D。再判斷問題2的答案，若肯定就選擇D，否則選A。

如果問題（1）的答案是否定的，那麼可能的選擇答案是B、C、E再判斷問題（2）的答案，若肯定就選擇B，否則有兩種可能的答案即C或E。

最後，再判斷問題（2）的答案，若肯定選擇C，否則選擇E。

採用這種解題方法，即使不能全部回答上述三個問題。也可以用來排除其中不可能或錯誤的選項。例如，如果你僅知道問題（1）的答案是肯定的，那麼你就能排除掉選項B、C和E；如果你僅知道問題（3）的答案是肯定的，那麼，你就能排除掉選項E；如果你僅知道問題2的答案是否定的，那麼你就能排除選項D和B。

只要考生能熟悉地掌握上述解題方法，那麼在數據填充部分中獲得高分並非難事。

3.2 Data Sufficiency Test: Question

1. In which year was Rahul born?

    (1) Rahul at present is 25 years younger to his mother.

    (2) Rahul's brother, who was born in 1964, is 35 years younger to his mother.

    A. I alone is sufficient while II alone is not sufficient

    B. II alone is sufficient while I alone is not sufficient

    C. Either I or II is sufficient

    D. Neither I nor II is sufficient

    E. Both I and II are sufficient

2. What will be the total weight of 10 poles, each of the same weight?

    (1) One-fourth of the weight of each pole is 5 kg.

    (2) The total weight of three poles is 20 kilograms more than the total weight of two poles.

    A. I alone is sufficient while II alone is not sufficient

    B. II alone is sufficient while I alone is not sufficient

    C. Either I or II is sufficient

    D. Neither I nor II is sufficient

    E. Both I and II are sufficient

3. How many children does M have?

   (1) H is the only daughter of X who is wife of M.

   (2) K and J are brothers of M.

   A. I alone is sufficient while II alone is not sufficient
   B. II alone is sufficient while I alone is not sufficient
   C. Either I or II is sufficient
   D. Neither I nor II is sufficient
   E. Both I and II are sufficient

4. How much was the total sale of the company?

   (1) The company sold 8000 units of product A each costing Rs. 25.

   (2) This company has no other product line.

   A. I alone is sufficient while II alone is not sufficient
   B. II alone is sufficient while I alone is not sufficient
   C. Either I or II is sufficient
   D. Neither I nor II is sufficient
   E. Both I and II are sufficient

5. The last Sunday of March, 2006 fell on which date?

   (1) The first Sunday of that month fell on 5th.

   (2) The last day of that month was Friday.

   A. I alone is sufficient while II alone is not sufficient
   B. II alone is sufficient while I alone is not sufficient

C. Either I or II is sufficient

D. Neither I nor II is sufficient

E. Both I and II are sufficient

6. What is the code for 'sky' in the code language?

(1)In the code language, 'sky is clear' is written as 'de ra fa'.

(2) In the same code language, 'make it clear' is written as 'de ga jo'.

A. I alone is sufficient while II alone is not sufficient

B. II alone is sufficient while I alone is not sufficient

C. Either I or II is sufficient

D. Neither I nor II is sufficient

E. Both I and II are sufficient

7. How many children are there between P and Q in a row of children ?

(1) P is fifteenth from the left in the row.

(2) Q is exactly in the middle and there are ten children towards his right.

A. I alone is sufficient while II alone is not sufficient

B. II alone is sufficient while I alone is not sufficient

C. Either I or II is sufficient

D. Neither I nor II is sufficient

E. Both I and II are sufficient

8. How is T related to K?

   (1) R's sister J has married Ts brother L, who is the only son of his parents.

   (2) K is the only daughter of L and J.

   A. I alone is sufficient while II alone is not sufficient

   B. II alone is sufficient while I alone is not sufficient

   C. Either I or II is sufficient

   D. Neither I nor II is sufficient

   E. Both I and II are sufficient

9. How is J related to P?

   (1) M is brother of P and T is sister of P.

   (2) P's mother is married to J's husband who has one son and two daughters.

   A. I alone is sufficient while II alone is not sufficient

   B. II alone is sufficient while I alone is not sufficient

   C. Either I or II is sufficient

   D. Neither I nor II is sufficient

   E. Both I and II are sufficient

10. How is X related to Y?

   (1) Y and Z are children of D who is wife of X.

   (2) R's sister X is married to Ys father.

   A. I alone is sufficient while II alone is not sufficient

   B. II alone is sufficient while I alone is not sufficient

C. Either I or II is sufficient

D. Neither I nor II is sufficient

E. Both I and II are sufficient

11. Who is to the immediate right of P among five persons P, Q, R, S and T facing North?

(1) R is third to the left of Q and P is second to the right of R.

(2) Q is to the immediate left of T who is second to the right of P.

A. I alone is sufficient while II alone is not sufficient

B. II alone is sufficient while I alone is not sufficient

C. Either I or II is sufficient

D. Neither I nor II is sufficient

E. Both I and II are sufficient

12. On which date of the month was Anjali born in February 2004?

(1) Anjali was born on an even date of the month.

(2) Anjali's birth date was a prime number.

A. I alone is sufficient while II alone is not sufficient

B. II alone is sufficient while I alone is not sufficient

C. Either I or II is sufficient

D. Neither I nor II is sufficient

E. Both I and II are sufficient

13. How is X related to Y?

    (1) Y says, "I have only one brother".

    (2) X says, "I have only one sister".

    A. I alone is sufficient while II alone is not sufficient

    B. II alone is sufficient while I alone is not sufficient

    C. Either I or II is sufficient

    D. Neither I nor II is sufficient

    E. Both I and II are sufficient

14. How is F related to P?

    (1) P has two sisters M and N.

    (2) F's mother is sister of M's father.

    A. I alone is sufficient while II alone is not sufficient

    B. II alone is sufficient while I alone is not sufficient

    C. Either I or II is sufficient

    D. Neither I nor II is sufficient

    E. Both I and II are sufficient

15. B is the brother of A. How is A related to B?

    (1) A is the sister of C.

    (2) E is the husband of A.

    A. I alone is sufficient while II alone is not sufficient

    B. II alone is sufficient while I alone is not sufficient

    C. Either I or II is sufficient

D. Neither I nor II is sufficient

E. Both I and II are sufficient

16. How many children are there in the row of children facing North?

(1) Vishakha who is fifth from the left end is eighth to the left of Ashish who is twelfth from the right end.

(2) Rohit is fifth to the left of Nisha who is seventh from the right end and eighteenth from the left end.

A. I alone is sufficient while II alone is not sufficient

B. II alone is sufficient while I alone is not sufficient

C. Either I or II is sufficient

D. Neither I nor II is sufficient

E. Both I and II are sufficient

17. How many doctors are practising in this town?

(1) There is one doctor per seven hundred residents.

(2) There are 16 wards with each ward having as many doctors as the number of wards.

A. I alone is sufficient while II alone is not sufficient

B. II alone is sufficient while I alone is not sufficient

C. Either I or II is sufficient

D. Neither I nor II is sufficient

E. Both I and II are sufficient

18. On which day of the week was birthday of Sahil?

    (1) Sahil celebrated his birthday the very next day on which Arun celebrated his birthday.

    (2) The sister of Sahil was born on the third day of the week and two days after Sahil was born.

    A. I alone is sufficient while II alone is not sufficient

    B. II alone is sufficient while I alone is not sufficient

    C. Either I or II is sufficient

    D. Neither I nor II is sufficient

    E. Both I and II are sufficient

19. How many pages of book X did Robert read on Sunday?

    (1) The book has 300 pages out of which two-thirds were read by him before Sunday.

    (2) Robert read the last 40 pages of the book on the morning of Monday.

    A. I alone is sufficient while II alone is not sufficient

    B. II alone is sufficient while I alone is not sufficient

    C. Either I or II is sufficient

    D. Neither I nor II is sufficient

    E. Both I and II are sufficient

20. How is Tanya related to the man in the photograph?

(1) Man in the photograph is the only son of Tanya's grandfather.

(2) The man in the photograph has no brothers or sisters and his father is Tanya's grandfather.

A. I alone is sufficient while II alone is not sufficient

B. II alone is sufficient while I alone is not sufficient

C. Either I or II is sufficient

D. Neither I nor II is sufficient

E. Both I and II are sufficient

21. Among T, V, B, E and C, who is the third from the top when arranged in the descending order of their weights?

(1) B is heavier than T and C and is less heavier than V who is not the heaviest.

(2) C is heavier than only T.

A. I alone is sufficient while II alone is not sufficient

B. II alone is sufficient while I alone is not sufficient

C. Either I or II is sufficient

D. Neither I nor II is sufficient

E. Both I and II are sufficient

22. Which word in the code language means 'flower'?

    (1)'de fu la pane' means 'rose flower is beautiful' and 'la quiz' means 'beautiful tree'.

    (2) 'de la chin' means 'red rose flower' and 'pa chin' means 'red tea'.

    A. I alone is sufficient while II alone is not sufficient

    B. II alone is sufficient while I alone is not sufficient

    C. Either I or II is sufficient

    D. Neither I nor II is sufficient

    E. Both I and II are sufficient

23. How many students in a class play football?

    (1) Only boys play football.

    (2) There are forty boys and thirty girls in the class.

    A. I alone is sufficient while II alone is not sufficient

    B. II alone is sufficient while I alone is not sufficient

    C. Either I or II is sufficient

    D. Neither I nor II is sufficient

    E. Both I and II are sufficient

24. Who is C's partner in a game of cards involving four players A, B, C and D?

    (1) D is sitting opposite to A.

    (2) B is sitting right of A and left of D.

A. I alone is sufficient while II alone is not sufficient

B. II alone is sufficient while I alone is not sufficient

C. Either I or II is sufficient

D. Neither I nor II is sufficient

E. Both I and II are sufficient

25. On a T.V. channel, four serials A, B, C and D were screened, one on eacn day, on four consecutive days but not necessarily in that order. On which day was the serial C screened?

(1) The first serial was screened on 23rd, Tuesday and was followed by serial D.

(2) Serial A was not screened on 25th and one serial was screened between serials A and B.

A. I alone is sufficient while II alone is not sufficient

B. II alone is sufficient while I alone is not sufficient

C. Either I or II is sufficient

D. Neither I nor II is sufficient

E. Both I and II are sufficient

26. How is Sulekha related to Nandini?

(1) Sulekha's husband is the only son of Nandini's mother.

(2) Sulekha's brother and Nandini's husband are cousins.

A. I alone is sufficient while II alone is not sufficient

B. II alone is sufficient while I alone is not sufficient

C. Either I or II is sufficient

D. Neither I nor II is sufficient

E. Both I and II are sufficient

27. Can Ritesh retire from office X in January 2006, with full pension benefits?

(1) Ritesh will complete 30 years of service in office X in April 2000 and desires to retire.

(2) As per office X rules, an employee has to complete minimum 30 years of service and attain age of 60. Ritesh has 3 years to complete age of 60.

A. I alone is sufficient while II alone is not sufficient

B. II alone is sufficient while I alone is not sufficient

C. Either I or II is sufficient

D. Neither I nor II is sufficient

E. Both I and II are sufficient

28. What is the code for 'or' in the code language?

(1) 'nik sa te' means 'right or wrong', 'ro da nik' means 'he is right' and 'fe te ro' means 'that is wrong'.

(2) 'pa nik la' means 'that right man', 'sa ne pa' means 'this or that' and 'ne ka re' means 'tell this there'.

A. I alone is sufficient while II alone is not sufficient

B. II alone is sufficient while I alone is not sufficient

C. Either I or II is sufficient

D. Neither I nor II is sufficient

E. Both I and II are sufficient

29. Madan is elder than Kamal and Sharad is younger than Arvind. Who among them is the youngest?

    (1) Sharad is younger than Madan.

    (2) Arvind is younger than Kamal.

    A. I alone is sufficient while II alone is not sufficient

    B. II alone is sufficient while I alone is not sufficient

    C. Either I or II is sufficient

    D. Neither I nor II is sufficient

    E. Both I and II are sufficient

30. On which date in August was Kapil born?

    (1) Kapil's mother remembers that Kapil was born before nineteenth but after fifteenth.

    (2) Kapil's brother remembers that Kapil was born before seventeenth but after twelfth.

    A. I alone is sufficient while II alone is not sufficient

    B. II alone is sufficient while I alone is not sufficient

    C. Either I or II is sufficient

    D. Neither I nor II is sufficient

    E. Both I and II are sufficient

3.3 Data Sufficiency Test: Answer

1. E

From both I and II, we find that Rahul is (35 - 25) = 10 years older than his brother, who was born in 1964. So, Rahul was born in 1954.

2. C

From I, we conclude that weight of each pole = (4x5) kg = 20 kg.

So, total weight of 10 poles = (20 x 10) kg = 200 kg.

From II, we conclude that:

Weight of each pole = (weight of 3 poles) - (weight of 2 poles) = 20 kg.

So, total weight of 10 poles = (20 x 10) kg = 200 kg.

3. D

From I, we conclude that H is the only daughter of M. But this does not indicate that M has no son. The information given in II is immaterial.

4. E

From I, total sale of product A = Rs. (8000 x 25) = Rs. 200000.

From II, we know that the company deals only in product A.

This implies that sale of product A is the total sale of the company, which is Rs. 200000.

5. C

From I, we conclude that 5th, 12th, 19th and 26th of March, 2006 were

Sundays.

So, the last Sunday fell on 26th.

From II, we conclude that 31st March, 2006 was Friday. Thus, 26th March, 2006 was the last Sunday of the month.

6. D

The only word common to I and II is 'clear' and as such, only the code for 'clear' can be ascertained from the given information.

7. E

From II, Q being in the middle, there are 10 children to his right as well as to his left. So, Q is 11th from the left. From I, P is 15th from the left.

Thus, from both I and II, we conclude that there are 3 children between P and Q.

8. E

From I, we know that L is T's brother and J's husband. Since L is the only son of his parents, T is L's sister.

From II, we know that K is L's daughter.

Thus, from I and II, we conclude that T is the sister of K's father i.e. T is K's aunt.

9. B

From II, we know that P's mother is married to J's husband, which means that J is P's mother.

10. C

From I, we conclude that Y is the child of D who is wife of X i.e. X is Y's father.

From II, X is married to Y's father. This implies that X is Y's mother.

11. C

From I, we have the order: R, -, P, Q.

From II, we have the order: P, Q, T.

Clearly, each one of the above two orders indicates that Q is to the immediate right of P.

12. E

From I and II, we conclude that Anjali was born in February 2004 on a date which is an even prime number. Since the only even prime number is 2, so Anjali was born on 2nd February, 2004.

13. D

The statements in I and II do not provide any clue regarding relation between X and Y.

14. E

From I and II, we conclude that P is M's brother and so M's father is P's father. So, F is the child of the sister of P's father i.e. F's mother is P's aunt or F is P's cousin.

15. C

B is A's brother means A is either brother or sister of B. Now, each one of I and II individually indicates that A is a female, which means that A is B's sister.

16. C

Since 8th to the left of 12th from the right is 20th from the right, so from I, we know that Vishakha is 5th from left and 20th from right i.e. there are 4 children to the left and 19 to the right of Vishakha. So, there are (4 + 1 + 19) i.e. 24 children in the row.

From II, Nisha is 7th from right and 18th from left end of the row.

So, there are (6 + 1 + 17) = 24 children in the row.

17. B

From I, total number of doctors in town = (1/700 x N) , where N = total number of residents in town. But, the value of N is not known.

From II, total number of doctors in town

= (Number of wards in town) x (Number of doctors in each ward)

= 16 x 16 = 256.

18. B

I does not mention the day of the week on the birthday of either Arun or Sahil.

According to II, Sahil's sister was born on Wednesday and Sahil was born two days before Wednesday i.e. on Monday.

19. E

From I and II, we find that Robert read (300 x 2/3) i.e. 200 pages before Sunday and the last 40 pages on Monday.

This means that he read [300 - (200 + 40)] i.e. 60 pages on Sunday.

20. B

From I, we conclude that the man is the only son of Tanya's grandfather i.e. he is Tanya's father or Tanya is the man's daughter.

From II, we conclude that the man's father is Tanya's grandfather. Since the man has no brothers or sisters, so he is Tanya's father or Tanya is the man's daughter.

21. A

From I, we have: B > T, B > C, V > B. Thus, V is heavier than each one of B, T and C. But V is not the heaviest. So, E is the heaviest.

Thus, we have the order. E>V>B>T>C or E>V>B>C>T. Clearly, B is third from the top.

22. D

From the two statements given in I, the code for the only common word 'beautiful' can be determined.

From the two statements given in II, the code for the only common word 'red' can be determined.

In I and II, the common words are 'rose and 'flower' and the common code words are 'de' and 'la'. So, the code for 'flower' is either 'de' or 'la'.

23. D

It is not mentioned whether all the boys or a proportion of them play football.

24. C

Clearly, each of the given statements shows that B is sitting opposite to C or B is the partner of C

25. E

From I, we know that the serials were screened on 23rd, 24th, 25th and 26th.

Clearly, D was screened second i.e. on 24th, Wednesday.

From II, we know that one serial was screened between A and B.

So, A and B were screened first and third, i.e. on 23rd and 25th. But, A was not screened on 25th.

So, A was screened on 23rd and B on 25th. Thus, C was screened on 26th, Friday.

26. C

From I, we conclude that Sulekha is the wife of Nandini's mother's only son i.e. Nandini's brother. Thus, Sulekha is Nandini's sister-in-law.

From II, we conclude that Sulekha is the cousin of Nandini's husband, which implies that Sulekha is Nandini's sister-in-law.

27. E

Clearly, the facts given in I and II contain two conditions to be fulfilled to get retirement and also indicate that Ritesh fulfills only one condition out of them.

28. C

I. In 'right or wrong' and 'he is right', the common word is 'right' and the common code word is 'nik'. So 'nik' means 'right'. In 'right or wrong' and 'that is wrong', the common word is 'wrong' and the common code word is 'te'. So, 'te' means 'wrong'.

Thus, in 'right or wrong', 'sa' is the code for 'or'. II. In 'that right man' and 'this or that', the common word is 'that' and the common code word is 'pa'. So, 'pa' means 'that'. In 'this or that' and 'tell this there', the common word is 'this' and the common code word is 'ne'. So, 'ne' means 'this'. Thus, in 'this or that', 'sa' is the code for 'or'.

29. B

As given, we have: M > K, A > S.

From II, K > A. Thus, we have: M > K > A > S.

So, Sharad is the youngest. From I, M > S. Thus, we have: M>K>A>S or M>A>K>S or M>A>S>K.

30. E

From I, we conclude that Kapil was born on any one of the dates among 16th, 17th and 18th.

From II, we conclude that Kapil was born on any one of the dates among 13th, 14th, 15th and 16th.

Thus, from both I and II, we conclude that Kapil was born on 16th August.

PART IV NUMERICAL REASONING

Numerical Reasoning（數字推理）
基本考核的課題包括度量衡、圖形、
分數、數型、單位兌換、組合、排列、
方向、時間、數字的分類排列關係
結合、因數、倍數、四則運算、價錢、
統計、代數、周界面積、算式推斷、
基本運算和關係及運算符號等。

4.1 數學推理及例題解析

1. 湊整法

【例題】

1. 5.213+1.384+4.787+8.616的值：

    A. 20

    B. 19

    C. 18

    D. 17

2. 99×55的值：

    A. 5,500

    B. 5,445

    C. 5,450

    D. 5,050

3. 4/2 - 1/5 - 3/4 - 4/5 - 1/4的值：

    A. 1/2

    B. 1/3

    C. 0

    D. 1/4

4. 19999+1999+199+19的值：

    A. 22219

    B. 22218

    C. 22217

    D. 22216

【答案及解析】

1. 答案：A

本題是小數湊整。先分別將0.213+0.787=1及0.384+0.616=1，然後將5+1+4+8+2=20，故本題的正確答案為A。

2. 答案：B

這是道乘法湊整的題。假如直接將兩數相乘則較為費時間，假如將99湊為100，再乘以55，那就快多了，只專心算即可。但要記住，在得數5,500中還需要減去55才是最終的得數，不然馬馬虎虎選A就錯了。故本題正確答案為B。

3. 答案：C

這是道分數湊整的題，可先將(1/5+4/5)+(3/4+1/4)=2心算出來，然後將4/2=2心算出來，2-2=0。故本題正確答案為C。

4. 答案：D

此題可用湊整法運算，將每個加數後加1，即19999+1=20000，1999+1=2000，199+1=200，19+1=20，再將四個數相加得22220，最後再減去加上的4個1，即22220-4=22216。故本題正確答案為D。

2. 觀察尾數法

【例題】

1. 2768+6789+7897的值：

    A. 17454

    B. 18456

    C. 18458

    D. 17455

2. 2789-1123-1234的值：

   A. 433

   B. 432

   C. 532

   D. 533

3. 891×745×810的值：

   A. 73951

   B. 72958

   C. 73950

   D. 537673950

【答案及解析】

1. 答案：A

這道題假如直接運算，則需花費較多的時間。假如專心算，將其三個尾數相加，得24，其尾數是4。再看4個選項，B、C、D的尾數不是4，只有A符合此數。故本題的正確答案為A。

2. 答案：B

這是道運用觀察尾數法計算減法的題。尾數9-3-4=2，選項A、D可排除。那麼B、C兩個選項的尾數都是2，怎麼辦？可再觀察B、C兩選項的首數，因為2-1-1=0，還不能確定，再看第二位數，7-1-2=4，只有選項B符合。故本題的正確答案為B。

3. 答案：D

這道題首先要觀察尾數，三個尾數相乘，1×5×0=0，因此，將A、B選項排除。那麼C、D兩選項中如何選擇出對的一項呢？因為3個三位數相乘，至少得出6位數的積，假如3個首位數相乘之積大於10的話，最多可

得9位數的積。C選項只有5位數,所以被淘汰,而D選項是9位數,符合得數要求。故本題的正確答案為D。

3. 未知法

【例題】

1.　17 580÷15的值:

　　A. 1173

　　B. 1115

　　C. 1177

　　D. 未給出

2.　2014年聖誕黃金旅遊檔期,在美國實現的390億元的旅遊收入中,民航客運收入16億元,比2012年同期增長18.5%,鐵路客運收入11.4億元,比2002年同期增長13.5%。下列敍述正確的是:

　　A. 2014年與2012年聖誕黃金旅遊檔期間,美國民航與鐵路客運收入上大體持平。

　　B. 2014年聖誕黃金旅遊檔期間,美國民航與鐵路客運收入合計27億元。

　　C. 未給出

　　D. 2014年與2012年聖誕黃金旅遊檔期間的客運收入上,民航與鐵路相比增加率多5%。

3.　5067+2433-5434的值:

　　A. 3066

　　B. 2066

　　C. 1066

　　D. 未給出

【答案及解析】

1. 答案：D

這道除法題的被除數尾數是0，除數的尾數是5，因此，其商數的尾數必然是雙數，因四個選項中的A、B、C三項尾數皆為單數，所以都應排除，實際上沒有給出正確值。故本題的正確答案為D。

2. 答案：D

A選項是錯的，因為2014年民航與鐵路客運收入都增長10%以上。

B選項也是錯的，2014年聖誕黃金旅遊檔期間兩項收入合計為16+11.4=27.4（億元），而不同於2002年同期的27億元。以上兩項排除後，還應看看D選項是否正確，假如錯了，當然就選C。但本題中，民航與鐵路客運量相比，增加率為18.5%-13.5%=5%，D是正確的。可見C選項是起干擾作用的。故本題的正確答案為D。

3. 答案：B

此題的四個選項中，除D之外的A、B、C三個選項，其後三位數完全相同，只注重觀察首位數誰是正確的就可以了。5+2-5=2，D選項在這裡起干擾作用。故本題的正確答案為B。

4. 互補數法

【例題】

1.　$3840 \times 78 \div 192$ 的值：

    A. 1540

    B. 1550

    C. 1560

    D. 1570

2. 4689-1728-2272的值：

A. 1789

B. 1689

C. 689

D. 989

3. $840 \div (42 \times 4)$ 的值：

A. 5

B. 4

C. 3

D. 2

【答案及解析】

1. 答案：C

此題可以將3840÷192=20，78×20=1560，故本題的正確答案為C。

2. 答案：C

此題可先專心算將兩個減數相加，1728+2272=4000。然後再從被減數中減去減數之和，即4689-4000=689。故本題的正確答案為C。

3. 答案：A

此題可先將840÷42=20專心算得出，然後再將已去掉括號後的乘號變成除號，20÷4=5。故本題的正確答案為A。

5. 基準數法

【例題】

1. 1997 + 1998 + 1999 + 2000 + 2001的值：

   A. 9993
   B. 9994
   C. 9995
   D. 9996

2. 2863+2874+2885+2896+2907的值：

   A. 14435
   B. 14425
   C. 14415
   D. 14405

【答案及解析】

1. 答案：C

碰到這類五個數按一定規律排列的題，可用中間數即1999作為基準數，而題中的1997=1999-2，1998=1999-1，2000=1999+1，2001=1999+2，所以該題的和為1999×5 (1+2-2-1)=1999×5=9995。在這裡不必計算，可將湊整法使用上，1999×5=2000×5-5=9995。故本題的正確答案為C。

2. 答案：B

該題初看不那麼好找規律，但仔細分析後可見，每相鄰的兩個數之間的差為11，也可取中間數2885作為基準數。那麼2863=2885-22，2874=2885-11，2896=2885+11，2907=2885+22。所以，該題之和為2 885×5 (22+11-22-11)=2 885×5=2900×5-75=14425。故本題的正確答案為B。

6. 求等差數列的和

【例題】

1. 2+4+6…22+24的值：

    A. 153

    B. 154

    C. 155

    D. 156

2. 1+2+3…99+100的值：

    A. 5030

    B. 5040

    C. 5050

    D. 5060

3. 10+15+20…55+60的值：

    A. 365

    B. 385

    C. 405

    D. 425

【答案及解析】

1. 答案：D

求等差數列之和有個公式，即（首項+末項）×項數÷2，項數＝（末項-首項）÷公差+1。在該題中，項數＝（24-2）÷2+1=12，數列之和＝（2+24）×12÷2=156。故本題的正確答案為D。

2. 答案：C

該題看起來較為複雜，計算從1到100之和，假如用1+99=100，2+98=100等之法計算，那將費時費力，而用求等差數列之和的公式計算，很快便可出結果。即(100-1)÷1+1=99×1+1=100，那麼該數列之和即為(1+100)÷2×100=5050。故本題正確答案為C。

3. 答案：B

該題的公差為5，依前題公式，項數=(60-10)÷5+1=11，那麼該題的值即(10+60)÷2×11=35×11=385。故本題的正確答案為B。

7. 快速心算法

【例題】

1. 做一個彩球需用8種顏色的彩紙，問做同樣的4個彩球需用多少種顏色的彩紙？

   A. 32

   B. 24

   C. 16

   D. 8

2. 甲的年齡是乙年齡的1倍，乙是30歲，問甲是多少歲?

   A. 60

   B. 30

   C. 40

   D. 50

【答案及解析】

1. 答案：D

仍用8種顏色的彩紙，A起干擾作用，切莫中了出題人的圈套。故本題的

正確答案為D。

2. 答案：B

本題説的甲與乙實際上是同歲，即30歲，切莫將1倍視為多1倍，即60歲，那就中了出題人的圈套。故本題的正確答案為B。

8. 加「1」計算法

【例題】

1.　一條街長200米，街道兩邊每隔4米栽一棵核桃樹，問兩邊共栽多少棵核桃樹？

　　A. 50
　　B. 51
　　C. 100
　　D. 102

2.　在一個圓形池子邊上每隔2米擺放一盆花，池周邊共長80米，共需擺多少盆花？

　　A. 50
　　B. 40
　　C. 41
　　D. 82

【答案及解析】

1. 答案：D

本題假如選A、B或選C都不對，因為(200÷4+1)×2=102。應注重兩點：
一是每邊起始點要種1棵，這樣每邊就要種200÷4+1=51（棵）；

二是兩邊共種多少棵，還需乘2，即51×2=102（棵）。故本題正確答案為D。

種樹棵數或放花盆數＝總長÷間距1。

2. 答案：B

這道題因為池周邊是圓形的，長80米，第一盆既是開始放的一盆，同時又是最後的一盆，所以不用加1盆，80÷2=40（盆）。在一條沒有終端的圓形池邊種樹或放花的盆數＝總長÷間距。故本題的正確答案為B。

9. 減「1」計算法

【例題】

1. 馬先生家住在第5層樓，假如每層樓之間樓梯台階數都是16，那麼馬先生每次回家要爬多少個樓梯台階？

    A. 80

    B. 60

    C. 64

    D. 48

2. 劉先生的家在某樓四門棟2層與4層各有一套住房。每層樓梯的台階數都是18，那麼劉先生每次從4層的住房下到2層的住房，共需下多少個樓梯台階？

    A. 36

    B. 54

    C. 18

    D. 68

【答案及解析】

1. 答案：C

住在5層的住戶，因為1層不需要上樓梯，只需爬2至5層的樓梯台階就可以了。所以本題的答案為16×(5-1)=64。故本題的正確答案為C。

樓梯台階數=層間台階數×(層數-1)

2. 答案：A

因為劉先生只下了兩層的樓梯台階，可直接用(4-2)×18=36即可。故本題的正確答案為A。

10. 爬繩計算法

【例題】

1. 一架單槓上掛著一條4米長的爬繩，趙先生每次向上爬1米後又滑下半米來。問趙先生需幾次才能爬上單槓？

    A. 8次

    B. 7次

    C. 6次

    D. 5次

2. 青蛙在井底向上跳，井深6米，青蛙每次跳上2米，又滑下1米，問青蛙需幾次方可跳出？

    A. 7

    B. 6

    C. 5

    D. 4

【答案及解析】

1. 答案：B

此題假如選A就中了出題人的圈套，實應選7次。因為爬了6次後，已經上了3米。最後一次爬1米就到頭了，不再往下滑了。故本題正確答案為B。

2. 答案：C

本題的原理同前題，不能選B，因為前4次共跳上4米，第五次就跳出井來了。故本題的正確答案為C。

11. 餘數相加計算法

【例題】

1. 今天是星期二，問再過36天是星期幾？

   A. 星期一
   B. 星期二
   C. 星期三
   D. 星期四

2. 今天是星期一，從今天算起，再過96天是星期幾？

   A. 星期二
   B. 星期四
   C. 星期五
   D. 星期六

【答案及解析】

1. 答案：C

這類題的算法是，天數÷7的餘數當天的星期數，即36÷7=5餘1，1+2=3。故本題的正確答案為C。

2. 答案：D

本題算法同前題，96÷7=13餘5，5+1=6。故本題正確答案為D。

12. 月日計算法

【例題】

1.　假如今天是2015年的11月28日，那麼再過105天是2016年的幾月幾日?

　　A. 2016年2月28日

　　B. 2016年3月11日

　　C. 2016年3月12日

　　D. 2016年3月13日

2.　才過生日的荷小姐今年28歲，她說了，她長了這麼大，按公曆才過了六次生日，問她生在哪月哪日？

　　A. 3月2日

　　B. 1月31日

　　C. 2月28日

　　D. 2月29日

【答案及解析】

1. 答案：C

計算月日要記住幾條法則。一是每年的1、3、5、7、8、10、12這七個月是31天，二是每年的4、6、9、11這四個月是30天，三是每年的2月，假如年份能被4整除，則該年的2月是29天，假如該年的年份不能被4整除，則是28天。記住這些非凡的算法，到時按月日去推算即可。

具體到這一題，11月是30天，還剩2天，12月、1月是31天，2月是28天，那麼2+31+31+29=93（天），105-93=12（天），即3月12日。故本題正確答案為C。

2. 答案：D

荷小姐生在2月29日，因為四年才有一次生日可過，所以她出生以來只過了六次生日。故本題的正確答案為D。

13. 比例分配計算法

【例題】

1.　一個村的東、西、南、北街的總人數是500人，四條街人數比例為1:2:3:4，問北街的人數是多少？

　　A. 250

　　B. 200

　　C. 220

　　D. 230

2.　一條長360米的繩子，按2:3:4的比例進行分截，最短的一截是多長？

　　A. 60

　　B. 70

　　C. 80

　　D. 90

3.　一所學校一、二、三年級學生總人數450人，三個年級的學生比例為2:3:4，問學生人數最多的年級有多少人？

　　A. 100

　　B. 150

C. 200

D. 250

【答案及解析】

1. 答案：B

四條街總人數可分成1+2+3+4＝10（份），每份為50人。北街佔4份，50×4=200（人），故本題的正確答案為B。

2. 答案：C

原理同上題，一份長為：360÷(2+3+4)=40（米），最短的一截為40×2=80（米），故本題的正確答案為C。

3. 答案：C

解答這種題，可以把總數看作包括了234=9份，其中人數最多的肯定是佔4/9的三年級，所以答案是200人。

14. 倍數計算法

【例題】

1.　甲是乙的三倍，乙是丙的1/6，問甲是丙的幾分之幾？

　　A. 1/2

　　B. 1/3

　　C. 1/4

　　D. 1/5

2. 張先生的藏書14000冊，馬先生的藏書18000冊。假如張先生想將自己的藏書成為馬先生藏書的3倍，那麼，他還應購進多少冊書？

A. 30000

B. 40000

C. 45000

D. 50000

【答案及解析】

1. 答案：A

在此題中，甲=3乙，乙=1/6丙。因此，甲=3×1/6丙=1/2丙。故本題的正確答案為A。

2. 答案：B

本題比較簡單，可先將14000與18000兩數字的三個零省去，那麼18×3=54，再減去張先生現有的書的冊數，54-14=40，再加上省去的三個零，即40000冊。故本題的正確答案為B。

15. 年齡計算法

【例題】

1. 女童今年4歲，媽媽今年28歲，那麼，多少歲時，媽媽的年齡是她的3倍？

A. 10

B. 11

C. 12

D. 13

2. 今年父親是兒子年齡的9倍，4年後父親是兒子年齡的5倍。那麼，今年父子年齡分別是多少歲？

    A. 40，5

    B. 35，6

    C. 36，4

    D. 32，6

【答案及解析】

1. 答案：C

今年媽媽比女童大28-4=24（歲），當媽媽年齡是女童年齡的3倍時，媽媽的年齡比女童大3-1=2（倍），即24歲正好是女童當時年齡的2倍。據此可推導出，女童在24÷2=12（歲）時，媽媽的年齡是她的3倍。驗證一下，4+8=12，28+8=36。故本題正確答案為C。

2. 答案：C

此題從直觀就可得知答案。只有(36+4)÷(4+4)=5，其他三個數分別加4，皆不得5。其實，這道題的答案一目了然，題中一開始就說了「父親是兒子年齡的9倍」，四個選項中，只有C符合條件。故本題的正確答案為C。

16. 雞兔同籠計算法

【例題】

1. 一籠中的雞和兔共250條腿，已知雞的隻數是兔的3倍，問籠中共有多少隻雞？

    A. 50

    B. 75

    C. 100

    D. 125

2. 一段公路上共行駛106輛汽車和兩輪摩托車，他們共有344個車輪，問汽車與摩托車各有多少輛？

   A. 68，38
   B. 67，39
   C. 66，40
   D. 65，41

【答案及解析】

1. 答案：B

設雞的隻數為x，按腿計算，雞腿為2x，雞為兔隻數的3倍，即兔是雞的1/3，兔子是4條腿，兔子的腿數為1/3x×4，即2x1/3x×4=250，10/3x=250，x=75（隻）。故本題正確答案為B。

通用公式總結：

雞數＝（兔腳數×總頭數-總腳數）÷（兔腳數-雞腳數）

兔數＝（總腳數-雞腳數×總頭數）÷（兔腳數-雞腳數）

2. 答案：C

該題的四個備選答案，其輛數合計為106輛，但汽車是4個車輪，摩托車是2個車輪。在四個選項中，66×4+40×2=344（個）車輪。故本題正確答案為C。

17. 人數計算法

【例題】

1. 一車間女工是男工的90%，因生產任務的需要又調入女工
   15人，這時女工比男工多20%，問此車間男工有多少人？

   A. 150

   B. 120

   C. 50

   D. 40

2. 某劇團男女演員人數相等，假如調出8個男演員，調進6個
   女演員後，女演員人數是男演員人數的3倍，該劇團原有
   多少女演員？

   A. 20

   B. 15

   C. 30

   D. 25

【答案及解析】

1. 答案：C

求男工數，可設男工為N，已知女工是男工的90%，即女工為0.9N，則
有：15+N+0.9N=1.2N+N，算出來就是N=50，即男工為50人。故本題
的正確答案為C。

2. 答案：B

從題中可知，女演員調進6人後，女演員人數則是男演員調出8人後的3
倍。故可設原男女演員皆為x，即x+6=(x-8)×3，x=15。

所以，女演員原來是15人。故本題的正確答案為B。

18. 工程計算法

【例題】

1. 一件工程，A隊單獨做300天完成，B隊單獨做200天完成。那麼，兩隊合作需幾天完成？

    A. 120

    B. 125

    C. 130

    D. 135

2. 一個水池有兩根水管，一根進水，一根排水。假如單開進水管，10分鐘將水池灌滿，假如單開排水管，15分鐘把一池水放完。現在池子是空的，假如兩管同時開放，多少分鐘可將水池灌滿？

    A. 20

    B. 25

    C. 30

    D. 35

【答案及解析】

1. 答案：A

本題的基本公式為，工作總量(假設為1)÷工作效率＝工作時間，即 $1 \div (1/300 + 1/200) = 120$。故本題的正確答案為A。

2. 答案：C

公式基本同上，$1 \div (1/10 - 1/15) = 30$。故本題正確答案為C。

4.2臨場秘技

A. 一秒作出判斷

1. 要熟練運用規律

拿到題目以後，怎樣一眼就能大致判斷出這道題目含有什麼規律呢？這也是有章可循的。

做題目時，我們能夠在一秒之內做出的判斷，就是一個數列項數的多少和數字變化幅度的大小，包括備選答案的數字的大小。根據這些信息我們就可以基本知道這個數列含有某種規律。

比如，給出的數列項數較多，有6項以上，一般可以首先考慮運用交替、分組和組合拼湊規律等。如果項數少就3項，一般只能用乘方和組合拼湊。如果數字之間變化幅度比較大，呈幾何級增長，多半要用到乘法、二級等比和乘方規律。剩下的可以考慮用加減法、等差及變式和質數規律。

此外，還可以根據數字之間變化呈現的曲線來判斷。比如，如果數字變化呈平緩的一條線，一般用加減法；如果數字變化呈現的線條比較陡，或者斜率絕對值較大，可以考慮用乘法、二級等比和乘方等；如果呈現拋物線形態，可考慮用乘方、質數等；呈U型線可考慮用減法、除法和乘方等；如果大小變動呈波浪線，主要考慮交替和分組。

2. 行測數字推理的技巧

公務員考試中，數字推理是很重要的一部分，儘管它佔的分值不多，但它的影響很大。這樣的題目看似很簡單，當你做題之後，往往會陷入做之不出、欲罷不能的境地，大多數考生很難在給出的時間裡做出答案，一般要花費雙倍或更多的時間，對後面的答題一很有大的影響。

那麼，要如何在規定的時間內或者在較短的時間裡做出題目呢？

首先，要準確理解什麼是數字推理。常規題型是給出一個缺少一項的數列，這個數列含有某種規律，要求考生運用這種規律從四個備選答案中選出一個填到數列的空缺處。我們在答題時，首先就要找出數列中含有什麼規律，再按照這種規律從四個選項中選出答案。這裡需要注意的是，這個數列可能包含多種規律，哪一個規律能用呢？這還要根據四個備選項來確定。

其次，要善於總結規律。數字推理題的解題關鍵就在於找規律，它的計算量不大，找到規律後很快就能得出答案。關鍵是能不能把這些東西變成你自己的？

最好的選擇還是自己去總結，我建議在理解題型的基礎上去總結規律。題目給出的是數列，就是一些數字的排列，能含有什麼規律，無外乎兩個方面，一是從「數」上去總結，就是數字本身或數字之間含有某些規律。如：具有相同性質的數排在一起，

呈現為奇偶數、質數規律等，還可以根據數的運算關系來排列，呈現為加減法、乘除和乘方等規律。

二是從「列」上去著眼，按照數列的性質，呈現出等差、等比規律。還可以根據數列的排列形式，呈現出雙重交替、分組、組合拼湊以及圓圈等。具體規律名稱叫什麼這並不重要，只要你熟知能用就行了。掌握了這些基本規律之後，在此基礎上盡可能發揮你的想像力，思考一下這些基本題型還可以有哪些變化形式，你能夠變化引申的越多，你的勝算就越大。

第三，要熟練運用規律。拿到題目以後，怎樣一眼就能大致判斷出這道題目含有什麼規律呢？

這也是有章可循的。做題目時，我們能夠在一秒之內做出的判斷，就是一個數列項數的多少和數字變化幅度的大小，包括備選答案的數字的大小。根據這些信息我們就可以基本知道這個數列含有某種規律。

比如，給出的數列項數較多，有6項以上，一般可以首先考慮運用交替、分組和組合拼湊規律等。如果項數少就3項，一般只能用乘方和組合拼湊。如果數字之間變化幅度比較大，呈幾何級增長，多半要用到乘法、二級等比和乘方規律。剩下的可以考慮用加減法、等差及變式和質數規律。

此外，還可以根據數字之間變化呈現的曲線來判斷。比如，如果數字變化呈平緩的一條線，一般用加減法；如果數字變化呈

現的線條比較陡，或者斜率絕對值較大，可以考慮用乘法、二級等比和乘方等；如果呈現拋物線形態，可考慮用乘方、質數等；呈U型線可考慮用減法、除法和乘方等；如果大小變動呈波浪線，主要考慮交替和分組。

我們可以下面4道題目為例：

【例題】

1.　102，96，108，84，132，（ ）

　　A. 36

　　B. 64

　　C. 70

　　D. 72

2.　-2，-8，0，64，（ ）

　　A. -64

　　B. 128

　　C. 156

　　D. 250

3.　3，7，16，107，（ ）

　　A. 1707

　　B. 1704

　　C. 1086

　　D. 1072

【答案及解析】

1. 答案：A

102-6=96，92+2×6=108，108-4×6=84，84+8×6=132，所以132應該減去16×6=36，故答案是A。

2. 答案：D

可以看出給出的數字稍加變化都是一些數的乘方，分析一下可知是自然數1，2，3，4立方的各項，對應乘以另一個數列-2，-1，0，1所得，下一個應該是5的立方乘以2，即$(-1)^3$x(-2)，$(-2)^3$x(-1)，$(-3)^3$x0，$(-4)^3$x1，$(-5)^3$x2=250，得出答案是D。

3. 答案：A

同樣，這道題的四個選項也比較大，但可以看出這些數和一些數的乘方離得較遠。再看能不能用乘法呢？從前兩項直接是看不出的，但是我們發現16與107的積和1707相近，相差5，往前推發現，前兩項的積減去5就等於後一項，因此答案是A。

最後，在利用這些規律的時候，還必須掌握一些基本的數理知識。如100以內的質數，30以內的自然數的平方，10以內的自然數立方，尾數是5的數的平方的速算，以及一些整數整除的速算法則等，你只要把這些知識簡單的複習一下就可以了。再加上適當的訓練，還有什麼題目做不出來呢？畢竟出題的思路就這麼多。

B. 兩把金鑰匙

首先，從時間上來考慮，能力傾向測試平均做每道題的時間在幾十秒左右，時間是非常緊張的。如果能在數字推理的每道題目上節省半分鐘，那麼整個考試就可以節省出幾分鐘，幾分鐘對於能力測試來說，可以說是非常珍貴的時間了。

其次，從心理上來考慮，如果能在數字推理上，達到又對又快的順利解決掉數字推理，那麼考生在做後面的題目時，心理上是會放鬆的，而且答題也會越來越自信；相反，如果在數字推理上卡住了，有題目沒做出來，那麼在後邊的答題中肯定會惦記著前面的題目，從而導致考試的緊張情緒，自己的信心也會被削減，甚至由於分神導致一些低級的失誤，例如漏答題等等。

因此，數字推理不論從應考的戰術，還是應考的戰略上來講都是非常重要的。

下面談談在考場上快速突破數字推理題目的「三把金鑰匙」：

第一把金鑰匙：利用特殊數字

一些數字推理題目中出現的數距離一些特殊的數字非常近，這裡所指的特殊數字包括平方數、立方數，因此當出現某個整數的平方或者立方周圍的數字時，我們可以從這些特殊數字入手，進而找出原數列的規律。例如下面這道題目：

【例題】

1. 0，9，26，65，124，( )

    A. 165

    B. 193

    C. 217

    D. 239

【答案及解析】

1. 答案：C

當我們看到26，65，124時，應該自然的本能的聯想到27，64，125，因為27，64和125都是整數的方次，27是3的立方，64是4的立方也是8的平方也是2的6次方，125是5的立方，很明顯，我們應該把64看作4的立方，也就是該數列每一項加1或減1以後，成為一組特殊的數字，他們是整數的立方，具體的說，就是：0+1為1的立方，9-1為2的立方，26+1為3的立方，65-1為4的立方，124+1為5的立方，因此，所求項減1應等於6的立方，故所求項為217，因此該題選C。

第二把金鑰匙：九九乘法口訣

很多時候，我們可以僅僅利用九九乘法口訣就找出已給數字的規律。例如下面這道題目：

【例題】

1, 1, 8, 16, 7, 21, 4, 16, 2, ( )

    A. 10

    B. 20

    C. 30

    D. 40

【答案及解析】

答案：A

當我們看到8，16，7，21，4，16時，如果能意識到它們在九九乘法口訣中的地位，那麼我們也就找到了解這道題的突破口了：1/1=1，16/8=2，21/7=3，16/4=4，因此所求項除以2應等於5，故所求項為10，故選A。因此，在做數字推理題時，應該一邊讀題，一邊考慮這些已知的數是否在乘法口訣中出現過，以及它們之間的聯繫。

C. 三招秒殺數字推理

第一招：看趨勢

　　拿到題目以後，用2秒鐘迅速判斷數列中各項的趨勢，例如：是越來越大，還是越來越小，還是有大有小。通過判斷走向，找出該題的突破口。有規律找規律，沒有規律做差。

【例題】

1.　7，9，12，17，24，（ ）

　　A. 27

　　B. 30

　　C. 31

　　D. 35

2.　14 ，6 ，2 ，0 ，（ ）

　　A. -2

　　B. -1

　　C. 0

　　D. 1

【答案及解析】

1. 答案：D

本題屬於多級數列。先看趨勢，越來越大，規律不明顯，兩兩做差，得到質數數列2，3，5，7，（11），所以選擇D選項。

2. 答案：B

本題屬於多級數列。題目中的一先看趨勢，越來越小，也就是趨勢是遞減的，是一致的。對於這類遞減的數列，我們通常的做法是從相鄰兩項的差或做商入手，很明顯，這道題目不能從做商入手（因為14/6不是整數），那麼，我們就作差，相鄰兩項的差為8，4，2成等比數列，因此，0減去所求項應等於1，故所求項等於-1，所以選擇B選項。

利用數列的趨勢，可以迅速判斷出應該采取的方法，所以，趨勢就是旗幟，趨勢就是解題的命脈。

第二招：看特殊數字

比如質數、平方數、立方數等。一些數字推理題目中出現的數距離這些特殊的數字非常近，因此當出現某個整數的平方或者立方周圍的數字時，我們可以從這些特殊數字入手，進而找出原數列的規律。

【例題】

1. 61，59，53，47，43，（ ），37

A. 42

B. 41

C. 39

D. 38

2. 0，9，26，65，124，（ ）

    A. 186

    B. 199

    C. 215

    D. 217

【答案及解析】

1. 答案：B

本題屬於質數數列。遞減的質數數列，所以選擇B選項。

2. 答案：D

本題屬於冪次修正數列。當我們看到26，65，124時，應該自然的聯想到27，64，125，因為27，64和125都是整數的次方，27是3的立方，64是4的立方也是8的平方也是2的6次方，125是5的立方。很明顯，我們應該把64看作4的立方，也就是該數列每一項加1或減1以後，成為一組特殊的數字，他們是整數的立方，具體的説：（），故所求項為217，所以選擇D選項。

第三招：看倍數關係

具體解題時，看相鄰的項、或者隔項之間有沒有倍數關係。

【例題】

1. 24，12，36，18，54，（ ）

    A. 27

    B. 30

    C. 32

    D. 33

2. 8，16，7，21，4，16，2，()

   A. 10
   B. 20
   C. 30
   D. 40

【答案及解析】

1. 答案：A

本題屬於多級數列。相鄰項的倍數很明顯，24是12的2倍，12是36的1/3，36是18的2倍，18是54的1/3，所以接下來是27，所以選擇A選項。

2. 答案：A

本題屬於多級數列。當我們看到8，16，7，21，4，16時，相鄰項有倍數關係，不是連續的，而是兩個兩個分開，1/1=1，16/8=2，21/7=3，16/4=4，因此所求項除以2應等於5，故所求項為10，故選A。

因此，在做數字推理題時，應該一邊讀題，看趨勢找規律，看特殊數，看倍數。希望這三招對我們的複習有所幫助。

PART V

INTERPRETATION OF TABLES AND GRAPHS

Interpretation of Tables and Graphs 的測試題形多變，有 Bar Chart、Line Chart、Pie Chart 及 Table Chart，考生宜勤加練習。

5.1 Bar Chart

The following bar graph shows the Income and Expenditures (in million US $) of five companies in the year 2001. The percent profit or loss of a company is given by:

% Profit/Loss = (Income - Expenditure)/Expenditure x 100

Income and Expenditure (in million US $) of five companies in the year 2001.

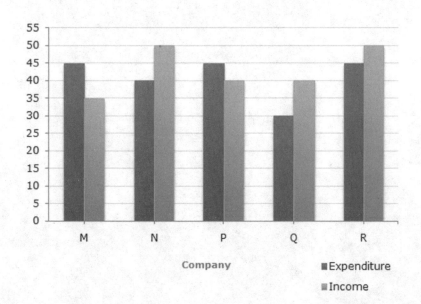

1. The companies M and N together had a percentage of profit/loss of?

    A. 12% loss

    B. 10% loss

    C. 10% profit

    D. There was no loss or profit

2. In 2001, what was the approximate percentage of profit/loss of all the five Companies taken together?

    A. about 5% profit

    B. about 6.5% profit

    C. about 4% loss

    D. about 7% loss

3. Which company earned the maximum percentage profit in the year 2001?

    A. M

    B. N

    C. P

    D. Q

4. For Company R, if the expenditure had increased by 20% in year 2001 from year 2000 and the company had earned profit of 10% in 2000, what was the Company's income in 2000 (in million US $)?

   A. 35.75

   B. 37.25

   C. 38.5

   D. 41.25

5. If the income of Company Q in 2001 was 10% more than its income in 2000 and the Company had earned a profit of 20% in 2000, then its expenditure in 2000 (in million US $) was?

   A. 28.28

   B. 30.30

   C. 32.32

   D. 34.34

Answer:

1. D

Total income of Companies M and N together

= (35 + 50) million US $

= 85 million US $

Total expenditure of Companies M and N together

= (45 + 40) million US $

= 85 million US $.

Percent Profit/Loss of companies M and N together

% Profit/Loss = (85-85)/85 x 100 = 0%

Thus, there was neither loss nor profit for companies M and N together.

2. A

Total income of all five companies

= (35 + 50 + 40 + 40 + 50) million US $

= 215 million US $.

Total expenditure of all five companies

= (45 + 40 + 45 + 30 + 45) million US $

= 205 million US $.

% Profit = 【(215-205)/205 x 100】% =4.88% = about 5%

3. D

The percentage profit/loss in the year 2001 for various comapanies are:

For M = 【(35-45)/45 x 100】%= -22.22% i.e., Loss = 22.22%

For N = 【(50-40)/40 x 100】%= 25% i.e., Profit = 25%

For P = 【(40-45)/45 x 100】%= -11.11% i.e., Loss = 11.11%

For Q = 【(40-30)/30 x 100】%= 33.33% i.e., Profit = 33.33%

For R = 【(50-45)/45 x 100】%= 11.11% i.e., Profit = 11.11%

Clearly, the Company Q earned the maximum profit in 2001.

4. D

Let the expenditure of Company R in 2000 be x million US $.

Then, expenditure of Company R in 2001 = (120/100 X X)million US $

so, 120x/100=45 》X= 37.5

i.e. expenditure of Company R in 2000 = 37.5 million US $

Let the income of Company R in 2000 be I million US $

Then, 10= (I-37.5)/37.5 x 100 (because % profit in 2000 = 10%)

》 I - 37.5 = 3.75

》 I = 41.25

i.e. Income of company R in 2000 = 41.25 million US $

5. B

Let the income of Company Q in 2001 = X million US $

Then, income of Company in 2001 = (110/100 X X)million US $

so 110X/100 = 40 》 X=400/11

i.e., income of company Q in 2000 = (400/11) million US $

Let the expenditure of Company Q in 2000 be E million US $,

Then, 20= 【(400/11)-E】 /E x 100 (because % profit = 20%)

》 20= 【(400/11E)-1】 x100

》 E=400/11 x 100/120 = 30.30

So, expenditure of company Q in 2000 = 30.30 million US $.

5.2 Table Chart

The following table gives the percentage distribution of population of five states, P, Q, R, S and T on the basis of poverty line and also on the basis of sex.

State	Percentage of Population below the Poverty Line	Proportion of Males and Females	
		Below Poverty Line	Above Poverty Line
		M : F	M : F
P	35	5 : 6	6 : 7
Q	25	3 : 5	4 : 5
R	24	1 : 2	2 : 3
S	19	3 : 2	4 : 3
T	15	5 : 3	3 : 2

1. If the male population above poverty line for State R is 1.9 million, then the total population of State R is?

   A. 4.5 million

   B. 4.85 million

   C. 5.35 million

   D. 6.25 million

2. What will be the number of females above the poverty line in the State S if it is known that the population of State S is 7 million?

   A. 3 million

   B. 2.43 million

   C. 1.33 million

   D. 5.7 million

3. What will be the male population above poverty line for State P if the female population below poverty line for State P is 2.1 million?

   A. 2.1 million

   B. 2.3 million

   C. 2.7 million

   D. 3.3 million

4. If the population of males below poverty line for State Q is 2.4 million and that for State T is 6 million, then the total populations of States Q and T are in the ratio?

   A. 1:3

   B. 2:5

   C. 3:7

   D. 4:9

Answer:

1. D

Let the total population of State R be x million.

Then, population of State R above poverty line

= [(100 - 24)% of x] million

=(76/100 x x)million

And so, male population of State R above poverty line

= 【2/5x(76/100)x x】 million

But, its given that male population of State R above poverty line=1.9 million.

So, 2/5 x 【(76/100)x x】 =1.9

》 X = (5x100x1.9)76x2 = 6.25

So, the total population of State R = 6.25 million.

2. B

Total population of State S = 7 million.

So, Population above poverty line

= [(100 - 19)% of 7] million

= (81% of 7) million

= 5.67 million.

And so, the number of females above poverty line in State S

=(3/7 x 5.67) million

=2.43 million

3. D

Female population below poverty line for State P = 2.1 million

Let the male population below poverty line for State P be x million.

Then, 5:6 = X:21 》 X=(2.1x5)/6 = 1.75

So, population below poverty line for State P = (2.1 + 1.75) million = 3.85 million.

Let the population above poverty line for State P by y million.

Since, 35% of the total population of State P is below poverty line, therefore, 65% of the total population of State P is above poverty line i.e., the ratio of population below poverty line to that above poverty line for State P is 35:65.

So, 35:65=3.85:y

》 y=(65 x 3.85)/35 = 7.15

So, population above poverty line for State P = 7.15 million and so, male population above poverty line for State P

$= (6/13 \times 7.15)$ million

$= 3.3$ million

4. B

For State Q:

Male population below poverty line = 2.4 million.

Let the female population below poverty line be x million.

Then, $3:5=2.4:x$ 》 $X=(5\times2.4)/3 = 4$

So, total population below poverty line = $(2.4+4)=6.4$ million

If Nq be the total population of State Q, then,

25% of Nq = 6.4 milion

》 $Nq =$ 【$(6.4 \times 100)/25$】 million = 25.6 million

For State T:

male population below poverty line = 6 million

Let the femaile population below poverty line be Y million

Then, $5:3=6:y$ 》 $y=(3\times6)/5 = 3.6$

So , total population below poverty line = $(6+3.6)$ = 9.6 million

If Nt be the total population of State T, then,

15% of Nt = 9.6 million

》 $Nt =$ 【$(9.6 \times 100)/15$】 million = 64 million

Thus, required ratio = Nq/Nt = 25.6/64 = 0.4 = 2/5

5.3 Pie Chart

Study the following graph and the table and answer the questions given below.

Data of different states regarding population of states in the year 2014

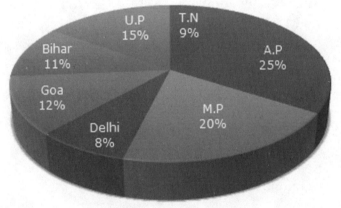

Total population of the given States = 3276000.

States	Sex and Literacy wise Population Ratio			
	Sex		Literacy	
	M	F	Literate	Illiterate
A.P	5	3	2	7
M.P	3	1	1	4
Delhi	2	3	2	1
Goa	3	5	3	2
Bihar	3	4	4	1
U.P.	3	2	7	2
T.N.	3	4	9	4

1. What will be the percentage of total number of males in U.P., M.P. and Goa together to the total population of all the given states?

    A. 25%

    B. 27.5%

    C. 28.5%

    D. 31.5%

2. What was the total number of illiterate people in A.P. and M.P. in 2014?

    A. 876040

    B. 932170

    C. 981550

    D. 1161160

3. What is the ratio of the number of females in T.N. to the number of females in Delhi?

    A. 7 : 5

    B. 9 : 7

    C. 13 : 11

    D. 15 : 14

4. What was the number of males in U.P. in the year 2014?

    A. 254650

    B. 294840

C. 321470

D. 341200

5. If in the year 2014, there was an increase of 10% in the population of U.P. and 12% in the population of M.P. compared to the previous year, then what was the ratio of populations of U.P. and M.P. in 2013?

   A. 42 : 55

   B. 48 : 55

   C. 7 : 11

   D. 4 : 5

Answer:

1. C

Number of males in U.P = [3/5 of (15% of N)]=3/5 x15/100 x N = 9 x N/100.

where N = 3276000.

Number of males in M.P =[3/4 of (20% of N)] = 3/4 x 20/100 x N = 15 x N/100

Number of males in Goa =[3/8 of (12% of N)] = 3/8 x12/100 x N = 4.5 x N/100

Therefore total number of males in these three states

= (9 + 15 + 4.5) x N/100

=(28.5 x N/100)

Therefore required percentage =   [(28.5 x N/100)/N x 100] % = 28.5%.

2. D

No. of illiterate people in A.P. = [7/9 of (25% of 3276000)]= 637000

No. of illiterate people in M.P. =[4/5 of (20% of 3276000)]= 524160

Therefore total number = (637000 + 524160) = 1161160

3. D

Required ratio

=[4/7of (9% of 3276000)]/[3/5of (8% of 3276000)]

=(4/7 x 9)/(3/5 x 8)

=(4/7 x 9 x5/3x 1/8)

=15/14

4. B

Number of males in U.P.

= [3/5 of (15% of 3276000)]

=3/5 x15/100 x 3726000

= 294840

5. A

Let x be the population of U.P. in 2013. Then,

Population of U.P. in 2014 = 110% of x =110/100 x x

Also, let y be the population of M.P. in 2013. Then,

Population of M.P. in 2014 = 112% of y =112/100 x y

Ratio of populations of U.P. and M.P. in 2014 =

(110/100 x x)/(112/100 x y)=110x/112y

From the pie-chart, this ratio is 15/20

Therefore, 110x/112y =15/20 => x/y = 15/20 x 112/110 = 42/55

Thus, ratio of populations of U.P. and M.P. in 2013 = x:y = 42:55

5.4 Line Chart

In a school the periodical examination are held every second month. In a session during April 2001 - March 2002, a student of Class IX appeared for each of the periodical exams. The aggregate marks obtained by him in each perodical exam are represented in the line-graph given below.

Marks Obtained by student in Six Periodical Held in Every Two Months During the Year in the Session 2001 - 2002.

Maximum Total Marks in each Periodical Exam = 500

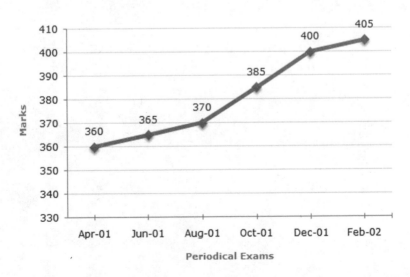

1. In which periodical exams did the student obtain the highest percentage increase in marks over the previous periodical exams?

   A. June, 01

   B. August, 01

   C. Oct, 01

   D. Dec, 01

2. The total number of marks obtained in Feb. 02 is what percent of the total marks obtained in April 01?

   A. 110%

   B. 112.5%

   C. 115%

   D. 116.5%

3. What is the percentage of marks obtained by the student in the periodical exams of August, 01 and Oct, 01 taken together?

   A. 73.25%

   B. 75.5%

   C. 77%

   D. 78.75%

4. What are the average marks obtained by the student in all the periodical exams during the last session?

    A. about 373

    B. about 379

    C. about 381

    D. about 385

5. In which periodical exams there is a fall in percentage of marks as compared to the previous periodical exams?

    A. None

    B. June, 01

    C. Oct, 01

    D. Feb, 02

Answer:

1. C

Percentage increase in marks in various periodical exams compared to the previous exams are:

For Jun 01 =[(365 - 360)/360 x 100]% = 1.39%

For Aug 01=[(370 - 365)/365 x 100]% = 1.37%

For Oct 01 =[(385 - 370)/370 x 100]% = 4.05%

For Dec 01 =[(400 - 385)/385 x 100]% = 3.90%

For Feb 02 =[(404 - 400)/400 x 100]% = 1.25%

Clearly, highest percentage increase in marks is in Oct 01.

2. B

Here it is clear from the graph that the student obtained 360, 365, 370, 385, 400 and 405 marks in periodical exams held in Apr 01, Jun 01, Aug 01, Oct 01, Dec 01 and Feb 02 respectively.

Required percentage =(405/360   x 100)% = 112.5%

3. B

Required percentage
=[(370+385)/(500+500)x 100]% = (755/1000x100)%= 75.5%

4. C

Average marks obtained in all the periodical exams
=1/6x [360 + 365 + 370 + 385 + 400 + 405] = 380.83 = about 381

5. A

As is clear from the graph, the total marks obtained in periodical exams, go on increasing. Since, the maximum marks for all the periodical exams are the same; it implies that the percentage of marks also goes on increasing.

Thus, in none of the periodical exams, there is a fall in percentage of marks compared to the previous exam.

看得喜 放不低

創出喜閱新思維

書名	投考公務員 題解EASY PASS 能力傾向測試（第四版）
ISBN	978-988-74806-0-0
定價	港幣$128
出版日期	2020年9月
作者	Fong Sir
責任編輯	投考公務員系列編輯部
版面設計	梁文俊
出版	文化會社有限公司
電郵	editor@culturecross.com
網址	www.culturecross.com
發行	香港聯合書刊物流有限公司
	地址：香港新界大埔汀麗路36號中華商務印刷大廈3樓
	電話：（852）2150 2100
	傳真：（852）2407 3062